QUARKS
Frontiers in Elementary Particle Physics

Alright Ruth, I about got this one renormalized — See Chapter 15

QUARKS

Frontiers in Elementary Particle Physics

Y. NAMBU

Translated by **R. Yoshida**

World Scientific
Philadelphia • Singapore

Published by

World Scientific Publishing Co Pte Ltd.
P. O. Box 128, Farrer Road, Singapore 9128
242, Cherry Street, Philadelphia PA 19106-1906, USA

Library of Congress Cataloging in Publication Data

Nambu, Y., 1921 –
  Quarks.

  Includes index.
  1. Particles (Nuclear physics)   2. Quarks.   I. Title.
QC793.2.N36   1985   539.7'21   85-5290
ISBN 9971-966-65-4
ISBN 9971-966-66-2 (pbk)

The publisher is grateful to Profs. R. Levi-Setti and Y. Yamaguchi
for providing photographs reproduced in this book.

Original Japanese language edition published by Kodansha Ltd., Tokyo,
and © 1981 Yoichiro Nambu.

Copyright © 1985 for the English edition by World Scientific
Publishing Co Pte Ltd.

Printed in Singapore by Singapore National Printers (Pte) Ltd.

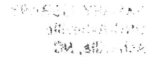

PREFACE

The purpose of elementary particle physics is to find the fundamental structure of matter and the laws that govern these objects. I have tried in this book to explain reasonably precisely, and also fairly completely, how the subject developed over the past 50 years and what is known today.

Fortunately, Japan has been making first-class contributions to particle physics. There probably is no one who does not know the names of Hideki Yukawa and Sin-itiro Tomonaga, but there have been many other scientists who have made important contributions to the field. These contributions are not just brilliant pieces of work but, as a whole, they have taught us the general directions which elementary particle theory should take.

Since I am also one of those who grew up under such a tradition, I have tried to present the process of thought by which physicists reached the present state of knowledge rather than presenting the newest knowledge as the orthodox view. Of course, doing this is impossible without assuming some specialized knowledge on the part of the reader, but I think it would be enough if the reader could get some idea of the research process from this book.

Elementary particle physics has made enormous progress in the last 50 years. Protons and mesons are no longer thought to be elementary; the quark has taken their place. Furthermore, there is a possibility of unifying the various forces previously thought to be unrelated. More astonishingly, the history of the Universe itself, the largest thing imaginable, has become completely intertwined with the problems of the smallest scale imaginable.

Of course, theory and experiment move hand-in-hand. Today's accelerators have energies over a million times that of old cyclotrons, require several thousand people to run and need huge government support. Still, researchers are dreaming about the next step. If we are fortunate and world peace prevails, these dreams will eventually become a reality. After contributing so much to the theoretical side, it would be good to see Japan contribute to the experimental side as well.

As I finished this book, Dr. Yukawa passed away. It is as though an era has gone by with him. I would feel fortunate if this book serves as an introduction for young readers who are interested in following the developments in elementary particle physics. I have made minor changes to the original text in order to bring this English edition up to date.

Finally I would like to thank Prof. Riccardo Levi-Setti and Dr. Taiji Yamanouchi for providing me with some material for the illustrations in this book.

Chicago
February, 1985   Yoichiro Nambu

CONTENTS

1

WHAT IS AN
ELEMENTARY PARTICLE?

Questions that may not have Answers

Most people, even those particularly interested in physics, have heard the term "elementary particle" somewhere. For example, an editorial column of the Asahi newspaper is entitled "elementary particle". Of course there is no relationship between the "elementary particle" newspaper column and the elementary particles we speak of in physics; yet such a title must have been chosen because it is a curiously attractive term. Why is it so attractive? This must be because the tradition of Japanese elementary particle theory, beginning with Hideki Yukawa's meson theory, is perceived by the people as something of which to be proud.

The Yukawa theory was born when I was a high school student. The term elementary particle had already begun to be used widely when I had begun to major in physics at my university; and thus several of my fellow students and I went to our professor and told him of our desire to study elementary particle theory.

In chemistry the word "element" is used to signify the basic building blocks of chemical substances; "element" implies that everything is made from it but nothing makes it up. Perhaps this kind of theory is born naturally when Man begins to think about his environment; but the modern scientific concept of elements is influenced heavily by the natural philosophy of the ancient Greeks.

What, then, is an element exactly? This question can be expressed in two parts. First, is there indeed such a thing as an element? And second, if there is such a thing as an element, what is it? It must be obvious by now that the aim of elementary particle physics is to answer these questions.

But the answer is not so easy. One doesn't even know if there really is an answer. Before the modern natural sciences there was a theory that earth, water, air, and fire were the elements. This theory was put forward by Aristotle; in English they still say that the influence of the natural world on Man, especially the damage done by rain, thunder, earthquake, and so forth, is caused by the "elements".

Of course, there is no one today who thinks that earth, water, air, and fire are the true elements. The philosophy of natural sciences which demand verification would force us to actually take matter and divide it until we reached the smallest constituents. But even if we did this literally we would not get very far. If we cut matter with a knife and examined matter with a microscope, we would never reach anything like a basic constituent. Particles about 1 micron in size still have shapes and seem to have internal structure; but we can not get to the next smaller scale using knives or microscopes. After all, the knife itself is made of elements, and therefore there is no way to make a knife sharper than the elements themselves. Since we need the size of the knife edge to be smaller than the things we wish to examine, we are back to the same question.

It was not until the 19th century when the science of chemistry made rapid progress that the idea of an atom was born. The ancient Greek philosopher Democritus theorized that all matter was made up of unchanging elements called atoms (atom = a-tom = not-divide, i.e., not divisible), but the atoms we speak of in chemistry were naturally conceived from the study of chemical reactions. This process of discovery was nothing like the naive notion of mechanical division I spoke of earlier, but was based on fundamentally different principles. We turn next to these principles.

Something Unchanging

If we mix some chemical substances A, B, C, D, . . . chemical reactions will take place. For example:

$$A + B = C, \tag{a}$$

$$A + D = E + F, \tag{b}$$

and so on. We want to find out the laws that govern the results of mixing these substances; to that end, let us see if there are any patterns in these reactions.

Let us say that we found, as the result of our examination, that the masses of the substances remain the same before and after the reactions. That is to say, if we change the interpretation of the above equations and think of A, B, etc., as the masses of substances A, B, etc., those equations become ordinary mathematical equations.

Here we have discovered one conservation law of nature — the law of conservation of mass. Substances may react together and change their character completely, but there is some quantity remains unchanged; this is the meaning of a conservation law.

Now let us suppose that after a more detailed study of the above reactions we learned the following: In order for the reaction (a) to take place, A and B must be mixed in the 1 to 2 ratio by mass. For example, if we try to mix equal quantities of A and B by mass, half of A is always left over. In the same way, the substances in the reaction (b) also have fixed mass ratios among A, B, E, and F. In other words, the masses of the substances that participate in the reactions have integer ratio relationships.

Thinking about how the above situation arises is nothing other than tracing the development of chemistry in the 19th century. Everybody knows the answers to these problems now, but it is easy to imagine the following train of thought. If there is something (substance) that can be represented by the smallest integer in many chemical reactions, then aren't all other substances made up of these smallest units? Doesn't the fact that the ratios are integers indicate that the substances in these reactions are themselves made of such fundamental units? Then let us name this fundamental unit "atom" and try to apply our new theory to other situations . . .

Chemistry, in reality, is much more complex. There is not just one kind of atom; the hydrogen atom is the lightest, but there are many heavier atoms (elements) that seem to have masses in integer multiples of the hydrogen mass, yet cannot be broken up into hydrogen atoms. If we study more closely, the integer multiple relationships of the atomic masses are not exact; even the conservation of mass is not strictly adhered to. These and many other problems begin to arise.

Is It Real?

What about the question of the existence of these atoms? There is no doubt in the mind of the man who does chemical experiments that there are such things as elements that exist as matter. Hydrogen, nitrogen, carbon, etc., can be actually isolated. But the atoms are a different story; we can't even tell the size of one atom. We couldn't possibly look at it with our eyes — we don't seem to be very far from Democritus' idea of the atom.

But there are big differences. First of all the chemical idea of the atom is based on a quantitative law and can be verified by experiment. Secondly, as research progresses, the existence of the atom begins to be proven from many different directions. We can find the size of the atom, and in a certain sense, we can see the atom with our own eyes.

For example, the tracks left by particles in a bubble chamber give the "illusion" that we can actually see the atoms. A man who was taught that the size of the atom is about 10^{-8} cm may immediately conclude that he couldn't possibly see an atom and the tracks must be an illusion; but is this correct?

Although we can see this book in front of us, we can't simply say that this book exists because it is large. What enters our brains is the stimuli of the light that is reflected by the book and not the book itself. When a charged particle enters a bubble chamber it excites and ionizes the liquid of the chamber; these ions become seeds for the bubbles and when these bubbles become sufficiently large, they reflect light and become visible. Now the processes involved in seeing tracks in bubble chambers are somewhat more complex than those involved in seeing a book, but can we say that they are fundamentally different?

To answer the question "What is existence?", which has been asked by philosophers from time immemorial, we must begin by pondering naive thoughts such as those above. The attitude of the scientist about such questions is quite ordinary; it is a simple extension of everyday logic.

We would try to touch a book if we thought its existence suspect. If we are still in doubt, we might ask someone nearby if he can see the book. If we try all the tests that are conceivable and still find no contradictions, we would conclude that the book is real. Of course, having once convinced ourselves that the book exists, we become unconscious of such questions.

But try to think of a situation when we encounter something completely new — say a UFO. How would we act? Then we would follow the steps for confirmation that were outlined above. The reality of the atom, in the final analysis, is established in the same way. If we suppose that the atom has such and such characteristics, and the atom passes all the tests we could perform to ascertain its characteristics, then we would come to believe in the existence of the atom. As we increase the kinds of tests to perform, our knowledge of the characteristics of the atoms becomes more and more accurate. We can also change or add to our theory of the atom without giving up the idea of the existence of the atom. If our theory were not a good one, then some inconsistencies would soon appear and we would have to do some unseemly patching up of the theory. If these patchings were needed one after the other, then our theory would probably be wrong and we would be better off scrapping the entire theory. On the other hand, if we had picked the correct theory, then we would be able to solve one mystery after another. This is very much like solving a crossword puzzle; we start out from what seem to be easy spots and try out some words. Even if we seem to succeed in some spots, many inconsistencies may crop up. We change our words a little but we still don't succeed. Then suddenly we have an inspiration and the rest is almost automatic.

When a theory in physics reaches such a stage, we begin to believe in the theory as true and real. But since physics is not a closed subject but something that changes and grows, this kind of peaceful situation does not last very long. We reach a stage when there is a breach in the theoretical structure and the previous theory becomes useless. And here, we must repeat our labors once again; but we must also remember that the old theory that has withstood so many tests cannot be completely wrong. Since the old theory is not useful in the new situation, we must build a new structure that contains the old theory as something that is applicable only to special situations.

It is still true that the atoms are the chemical "elements"; but when I was a student it was believed that such things as the Yukawa mesons were the true elementary particles. Today even the general public knows of such terms as quarks and leptons. If you ask today's physicist, he would say that these were the true "fundamental particles".

What, then, are these quarks and leptons? What is their relationship with the atoms and mesons? It is the purpose of this book to answer these questions, and in doing so we will learn the history of the development of physics in this century.

But let us now stop telling the story in chronological order. Let us begin at the present and look back into the past.

2

QUARKS AND LEPTONS

The Quark — The Unconventional Elementary Particle that has not been Found

The oddly-named quark was thus christened by one of the originators of the quark theory, Murray Gell-Mann. I will speak of the origins of this particle in some detail later; the quark is one of the particles thought to be "elementary" at present, and it still has not escaped the realm of fiction completely.

As I have said before, the words "elementary particles" refer to the most fundamental constituents of matter. When I say that the quark has not escaped the realm of fiction, I am being very careful and conservative since the quark theory can explain all phenomena known at present. What I mean is that we have not reached the stage, as we have with protons and electrons for example, where we have absolutely no doubt about the existence of the quark.

Why is there this slight doubt? It is because, although the quark is supposed to have characteristics unseen in the known particles, no one has spotted anything that looks like it. It appears that one cannot take a quark out of matter and confirm its characteristics. On the contrary, other known particles can be taken out by themselves and measurements can be made on them. Nucleons and electrons were mere theoretical constructs based on conservation laws and so forth at first, but they were eventually isolated and and their masses and charges were measured.

Actually, the quark, if it existed by itself, should be very easy to identify. This is because the electric charge carried by the quark is supposed to be two-thirds or one-third of the unit charge, i.e., the charge carried by an electron or a proton. All elements known up to now, be they electrons or nuclei, have electric charges either zero or an integer multiple (± 1, ± 2, . . .) of the unit charge e carried by an electron. So no matter what piece of matter is examined, its total charge is always an integer multiple of e.

One other important thing is the law of conservation of charge. Although reactions occur among particles and one particle can change into another, or particles exchange their electric charges, the total number of charges never changes. This is a typical example of the conservation laws I mentioned above, and if this is true it is not hard to imagine an element that carries the smallest possible amount of charge. But there is not just one kind of element that carries the smallest charge. Electrons, protons and many others that differ in mass and other characteristics have the electric charge $\pm e$.

It is theorized that, against the norm, valid for all known particles, quarks have the unprecedented charge of $\pm e/3$ or $\pm 2e/3$. That is to any, the ultimate unit of charge is not e but $\pm e/3$. If that is the case, are those particles that carry charge $\pm e$ such as electrons and protons not elementary particles but compounds made of quarks?

The answer is this: The proton is indeed a compound particle made of three quarks. In fact the quark theory grew out of conceiving the proton to be a compound particle. On the other hand, the electron is not made of quarks and is still thought to be elementary.

Heavy Particles, Light Particles, In-between Particles

I stated before that the leptons and quarks are the fundamental particles; the electron is one of the leptons. The name "lepton" comes from Greek meaning light particle. Among the other leptons are neutrinos (ν) and muons (μ), but leptons other than electrons usually do not appear in everyday phenomena. In any case, the electric charge of a lepton is $\pm e$ or 0.

What is the "heavy" particle as opposed to the light lepton? The Greek word for it is "baryon", and both protons and neutrons belong to this group. As is well known, the proton, or the hydrogen nucleus, has a mass some 1,800 times that of the electron. The neutron (n) is also a baryon, and several protons and neutrons congregate to make up an atomic nucleus; the nucleus pulls a cloud of electrons around itself to make up a neutral atom.

Protons and neutrons, as they are the ingredients of the nucleus, are collectively called nucleons (N). Other than these nucleons, baryons include the lambda particle (Λ), sigma particle

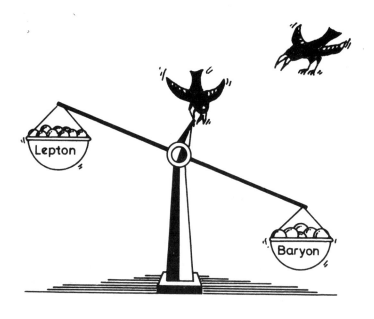

Baryon (heavy particle) and Lepton (light particle).

(Σ), and other unstable heavy particles. These baryons are now thought to be not fundamental particles but made of three quarks.

In between the baryons and the leptons, there exist "mesons" or "middle particles". These include the pion (π) whose existence was first predicted by Dr. H. Yukawa. As befits the name meson, the mass of the pion is 270 times that of the electron and one-seventh that of the proton.

However, classifying particles by their masses is not really very meaningful. There are many mesons that have masses comparable to those of baryons, and the muon, which is a lepton, has a mass about that of the pion. The recently discovered lepton, the tau (τ) has a mass larger than that of the proton.

The baryons and mesons together are called hadrons. This word means "strong particle" in Greek and these are thus named because they interact strongly. The force responsible for the

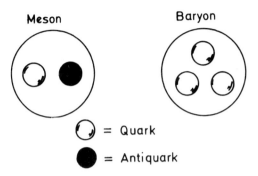

Fig. 2.1 Meson and Baryon contents.

strong interactions, or the strong force, is different from electromagnetic or gravitational forces, and the force that binds the
nucleus together, or the nuclear force, is one of its manifestations.
The force between two nucleons is thought to arise out of the
exchange of mesons among the nucleons; one emits a meson and
the other absorbs it. Since the exchange occurs frequently, the
nuclear force is stronger than the electromagnetic force.

According to the quark theory, baryons are made of three
quarks whereas the mesons are made of two quarks — actually,
one quark and one antiquark. This is the first time the word "anti"
has been mentioned; it is a prefix that can be applied to any lepton
or a hadron. For every particle there exists its "antiparticle". As
implied by its name, an antiparticle has opposite signed electric
charge and opposite signed other "quantum numbers" from those
of the particle, but they have the same mass. These quantum
numbers include numbers that characterize particles such as the
strangeness quantum number which will be discussed later. The
particle-antiparticle pair is like a pair of twins and they are not
completely different particles. The antiparticle of the electron,
the "antielectron", is usually called the "positron" and carries an
electric charge $+e$.

The reason we do not see positrons and antiprotons in everyday life is because when these meet up with their partners,
electrons and protons, the particle and the antiparticle combine

and disappear, or they "pair annihilate", and their energies appear as, for instance, photons. Conversely, if one wants to make an antiparticle, it has to be "pair produced" with its particle. Thus, we often do not distinguish between particles and antiparticles when counting the kinds of particles, and following this convention, we have been including antiquarks when discussing quarks in general.

Why don't the mesons pair annihilate if they are made of a quark and an antiquark? Part of the reason is that there are several kinds of quark, and mesons in general are not made of the same kind of quark pairs; but deeper arguments must be left until later.

Let's summarize the story so far. Hadrons ("strong particle") are a group of particles that interact strongly and are made of quarks. Among hadrons, the baryons ("heavy particles") are made of three quarks. (Antibaryons, therefore, are made of three antiquarks.) The mesons are made of two quarks (actually a quark and an antiquark). Other combinations do not seem to exist and, in particular, a single quark has not been observed.

Normally, the substances we observe are made, ultimately, of leptons and quarks. Quarks come together to make baryons, baryons come together to make nuclei, nuclei and electrons come together to become atoms, atoms come together to become molecules, molecules come together to become organisms and so on, . . .

The leptons ("light particles") include such relatively light particles as electrons and neutrinos which do not interact strongly, and these do exist by themselves.

Thus the structure of matter divides into several layers. This means that there are several scales of size and energy. The size scale is obvious. The size of an atom is about 10^{-8} cm, but if we look at its fine details, we will see electron clouds around a nucleus which is about 10^{-13} cm. Then if we look at the details of the nucleus, we will find protons and neutrons; and if we look at the details of the nucleons we "ought" to see the quarks. But when we begin to discuss what we mean exactly by size, all kinds of problems begin to arise.

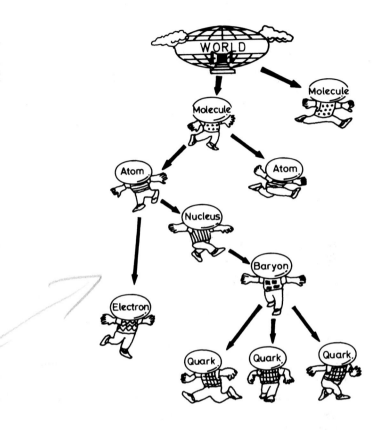

Layered structure of the material world.

What is the "Size" of a Particle?

One of the problems that arise is the extent of a particle as a "wave" as dictated by the principles of quantum mechanics. The wave, by its very nature, has an extent but has no set size. But the extent of a wave can be controlled by how one makes a wave. If one wants to confine a wave to a small volume, then it is necessary to choose one made up of short wavelength. Then by the de Broglie relationship which states that the wavelength is inversely proportional to the momentum of a particle, the uncertainty in momentum, and thus the uncertainty in the kinetic energy, becomes large. This is what is known as Heisenberg's uncertainty principle. Since particles with higher energy can be confined in smaller volume, it is necessary to achieve high energies in order to probe the internal structures of matter by striking it with particles.

Another problem is the range of the forces that act between the particles. Forces such as gravity and the Coulomb (electro-magnetic) force, whose strength varies as the inverse square of the distance, are said to have infinite range. On the other hand, a "Yukawa type" force such as the nuclear force has a finite range and weakens exponentially beyond a certain distance; you can actually think of these forces as not having any real effect beyond a certain range.

Since the range of the nuclear force is about 10^{-13} cm, and since the motion of two nucleons within the range of the nuclear force is strongly disturbed, it looks as if the actual nucleon size is about 10^{-13} cm. Since the strong force that acts between hadrons generally has a range of about 10^{-13} cm, all hadrons, regardless of type, may be said to have a size of about 10^{-13} cm.

Things to be Learned from Scattering Experiments

Many of the experiments that probe the structure of particles are like target shooting. Think of a particle as a "target" and think of another as a "bullet". Though the path of the bullet has a quantum mechanical uncertainty, this can be made arbitrarily small by raising the energy. But if the force between the target and the bullet is the finite-range strong force, the bullet will always "strike" the target if it comes within range of the force.

Typical quantities measured in a target shooting experiment are the "scattering cross section", i.e., size of the target, and the "angular distribution", i.e., the probability that the bullet is scattered into a particular direction. In order to measure the above quantities, we must have a constant stream (beam) of bullets of the same energy and a detector at a particular position to count the number of particles that come into that detector. By changing the angle of the detector placement and seeing the change in the number, the angular distribution is found, and by adding up the numbers measured at all angles and comparing the result with the incoming particle rate, the cross section is obtained.

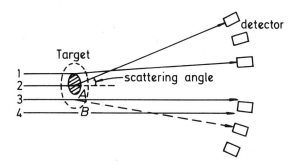

Fig. 2.2 The angle through which the bullet is deflected from its initial direction is the scattering angle. If the target is B which is larger then A than bullet number 3 which would not have been deflected by A would scatter as shown by the dotted line. Thus by counting the number of scattered bullets we can find the size (total cross section) of the target.

Of course, scattering takes place between an electron and a proton, too, but this is due to the long-range Coulomb force (Coulomb scattering) and differs from the scattering due to the strong force in such respects as the angular distribution and the variation with energy. In Coulomb scattering, the incoming particle often scatters at a place far away from the center of the target and deflects through only a small angle. (This tendency becomes more pronounced if the charge of the target is spread out. It is this fact that led Lord Rutherford to conclude that there is a small nucleus in the atom that carries positive charge and causes the large angle scattering that he observed.)

I have stated before that many kinds of reactions can happen besides simple scattering when two particles collide. Even in the Coulomb scattering of the electron and the proton, electromagnetic waves, or photons, may be emitted. If an electron collides with a hydrogen atom at an energy higher than a "threshold" value, the latter may be broken up into a proton and an electron.

One can write these reactions in equation form:

$$e + p \rightarrow e + p + \gamma,$$
$$\rightarrow e + p + \gamma + \gamma',$$
$$e + H \rightarrow p + e + e'.$$

Here, e is an electron, p is a proton, γ is a photon (gamma ray), and, of course, H is a hydrogen atom.

These reactions are called "inelastic scattering" as opposed to "elastic scattering". One of the characteristics of reactions between two hadrons is that both the probability and the kinds of inelastic scattering are large. That is to say, if A and B collide, all kinds of particles can come out, say, A, B, C, D, . . . And no matter which particles we choose, the range of the force is about the same and the frequency of the reaction is also about the same.

Thus the character of the hadron is both simple and complex. The fact that there are many kinds of hadrons suggest that they are not elementary but have some internal structure; besides, it does not appear that some of these hadrons are elementary while others are not.

Therefore, the first thing to do in understanding hadron reactions is to find the conservation law I spoke about in the previous chapter. Then after finding order in the hadron reactions, we hypothesize appropriate elementary particles that make up hadrons. This is exactly the same procedure as when atoms were hypothesized from the order found in chemical reactions.

In fact history took the above same path in the case of hadrons; this new conservation law is called the "Gell-Mann-Nakano-Nishijima law", and the new elementary particle is the "quark" conceived by M. Gell-Mann and G. Zweig.

3

LOOKING FOR QUARKS

Structure of Protons and Neutrons

As stated in the last chapter, the quarks are the elementary particles that make up hadrons. Of the hadron tribe, baryons are thought to be composed of three quarks and mesons of two quarks (one quark and one antiquark). The most unusual point about the quark theory, as M. Gell-Mann and G. Zweig independently formulated it, was that the quark carried a fractional charge. I intend to gradually explain why fractional charges were chosen, but let us begin now with the most common quarks u and d.

The above two letters correspond to "up" and "down", and these quarks are the constituents of protons and neutrons. Since all atomic nuclei are made of protons and neutrons, these two quarks can be said to be the fundamental constituents of all chemical elements. The proton has charge e and the neutron has charge zero. Therefore, if a nucleus is made of Z protons and N neutrons, the total charge of the nucleus would be Ze. The nucleus then captures Z electrons around itself and makes a neutral atom. Z is called that atom's atomic number; the combined total number of protons and neutrons, $Z+N=A$, is called the mass number. For example, the ordinary hydrogen atom has $Z=1$, $N=0$, $A=1$; heavy hydrogen has $Z=1$, $N=1$, $A=2$; ordinary helium has $Z=2$, $N=2$, $A=4$.

To obtain the total mass of the nucleus or the atom, one adds together the masses of the constituent particles and subtracts the binding energy, that is to say, the amount of energy released when the nucleus is formed, as expressed in mass units by Einstein's "energy equals mass" ($E=mc^2$) relation. But since the masses of the nucleons take up most of the mass of the nucleus and also since the masses of the proton and neutron are almost identical, the mass number $A=Z+N$ gives a rough estimate of the mass of the nucleus. (The mass of the electron is a mere $\frac{1}{1800}$ of that of the proton.) The chemical properties of an atom are determined by its electrons, and the mass of an atom is determined by A. Atoms which have the same Z but different A (or N) are called isotopes; hydrogen and heavy hydrogen are examples of isotopes.

As protons and neutrons do not differ very much aside from the electric charge, the u and d quarks are alike except for the electric charge.

If we apply atomic number and mass number to quarks, we have

$$u : Z = \tfrac{2}{3}, \qquad A = \tfrac{1}{3},$$
$$d : Z = -\tfrac{1}{3}, \qquad A = \tfrac{1}{3}, \tag{a}$$

and the proton and neutron are respectively expressed by the following chemical equations:

$$p = uud,$$
$$n = udd. \tag{b}$$

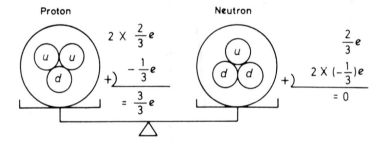

We can also check that if we start out with equations (b) and also assume that the masses of u and d quarks are about the same but that their charges differ by one, then we arrive at (a).

Why are three quarks used? The answer to this question will be deferred until later. But since we can take protons and neutrons out of nuclei, why can't we take nucleons and mesons apart into quarks? Why can't we see particles with electric charges $-\tfrac{1}{3}$ and $\tfrac{2}{3}$?

Looking for Fractionally Charged Particles

Finding the charge of a particle is relatively easy. When a charged particle goes through a cloud chamber or a bubble chamber, it knocks electrons off the atoms of the medium (gas or liquid) in the chamber; these electrons act like seeds of charged particles and form a track of cloud condensation or bubbles where the original charged particles went through. The thickness of this track is proportional to the square of the charge of the particle and is inversely proportional to the square of the speed of the particle; therefore high energy particles of charge $\pm e$ which move at nearly the speed of light all leave, more or less, the same thickness of the track. So if we find a thinner track than normal, the charge must be less than e.

Let us look for a quark using the above principle. We accelerate protons using a big proton accelerator (proton synchrotron) and send this beam of protons into a liquid hydrogen bubble chamber. Collisions between protons will produce various particles which leave tracks. Then we take a photograph of these tracks and look for one which is thinner than the normally minimum thickness track. But, in fact, no such tracks have been found.

A more sensitive method of looking for fractional charge is the oil drop experiment used by R.A. Millikan when he determined the size of the unit charge e. If one puts an oil drop into a liquid, it falls gently as is determined by the balance of gravity and friction. But the drop sometimes loses an electron to the liquid or gains an electron from the liquid. This gives the drop a positive or negative electric charge. So if an external electric field is applied, the drop will also feel an electric force and will change the direction of motion every time it changes charge. Millikan measured the motion of the oil drop and discovered that the change of electric charge occurred in units of e, which was a quantity he could measure.

What would happen if a free quark existed somewhere inside some matter? According to the quark theory, all nuclei contain quarks in multiples of three so there is no way a quark could combine with a nuclei or an electron to give integer charge. Even if the quark is unstable and decayed into some lighter particle, it couldn't decay just into integer charged particles. For example, if the d quark is heavier than the u quark, then the

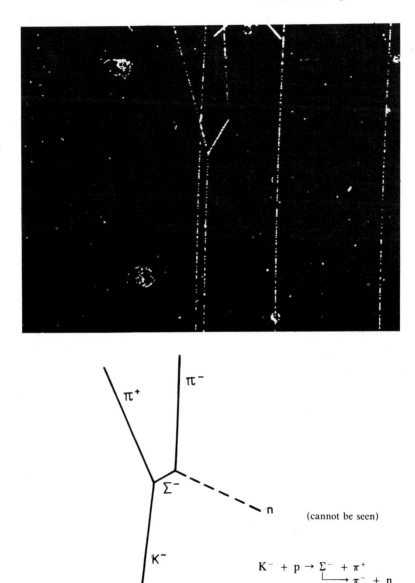

Fig. 3.2 Tracks of charged particles in a bubble chamber. This photograph shows a K^- beam reacting with protons and producing π^+ and Σ^+. The latter decays into π^- and n. (*Courtesy of Prof. Levi-Setti*)

same reaction as in beta decay, i.e., decay of the neutron into proton and an antineutrino, may occur. The equations for these reactions are:

$$n \rightarrow p + e + \bar{\nu},$$
$$d \rightarrow u + e + \bar{\nu},$$

both of which satisfy conservation of electric charge. But if there are no lighter quarks than u, u cannot change into anything else. This means that at least one of the quarks must be stable. It's the same story for antiquarks. Antiquarks and quarks can annihilate and turn into integer charged mesons, but if there is any quark or antiquark left over, these cannot be gotten rid of.

Suppose that there is a single u quark by itself in an oil drop, say, one millimeter in diameter. What would happen if such a drop existed in Millikan's experiment? As this drop gains and loses electrons, its charge would change to $\frac{2}{3}$, $\frac{2}{3} \pm 1$, $\frac{2}{3} \pm 2$, . . . which is clearly different from 0, ± 1, ± 2, ± 3, . . . which we normally expect.

Since the size of an atom is about 10^{-8} cm, an oil drop about 1 mm would have about $10^7 \times 10^7 \times 10^7 = 10^{21}$ atoms. This means we can find one free quark among 10^{21} atoms.

This Millikan-type experiment was carried out by W.M. Fairbank at Stanford University, and G. Morpurgo at the University of Genoa, among others. These modern experiments, however, do not use oil drops. Fairbank uses small spheres of niobium at very low superconducting temperatures. Since superconducting bodies repel magnetic fields completely, the niobium spheres will float in a magnetic field in a vacuum. If a sphere has an electric charge it will oscillate in an externally applied oscillating voltage, and its electric charge can be measured.

Morpurgo used ordinary iron spheres instead of superconductors, but the principle is the same.

What were their results? Some readers may remember reading in newspapers or magazines that Fairbank "discovered" the quark. He has made measurements over several years on samples of niobium spheres, and reports that he observed the charge of $\pm\frac{1}{3}$ in several of them. (As is probably clear from previous explanations, we cannot tell if it is a charge $+\frac{2}{3}$ u quark or a charge $-\frac{1}{3}$ d quark since the sphere can gain or lose electrons; the same holds for $-\frac{2}{3}$ $\bar{u}$ quark and $+\frac{1}{3}$ $\bar{d}$ quark.)

But it is still not clear if the above result is indicative of the existence of a free quark. The most mysterious point about the results of the above experiment is that the observed charge sometimes jumps from 0 to $\pm\frac{1}{3}$. This would not happen unless quarks were floating around all over the place just like electrons, frequently sticking to or getting off the sphere. But this would contradict the negative results of Morpurgo and others who found no fractional charges. Thus we must say that the existence of a free quark still has not been established.

4

VARIOUS ACCELERATORS

Profundity of Nature — Stronger Blows Produce Deeper Resonances

The symbol of elementary particle physics is the giant accelerator. Without this tool there can be no particle physics experiments, and without experiments there can be no progress in physics. Of course there are areas in elementary particle physics which do not require vast amounts of energy; experiments in these areas look for some subtle effects or require highly accurate measurements. But to discover new particles or investigate an unknown interaction, higher and higher energies are absolutely necessary. The "energy equals mass" relationship limits the mass of the particle that can be created at a given energy; so if we have looked at all the possible reactions with a given accelerator, we may say that the accelerator has fulfilled its fundamental purpose. Then we need an accelerator at the next higher energy — and so on. One might ask where is the guarantee that new discoveries will be made once we go to a higher energy. Frankly, there is no such guarantee. But from past experience, at least, there has not been a new accelerator which did not discover new particles. Even though we started out by expecting a small number of fundamental elements, nature surprised us by coming out with more and more new particles. Sometimes, of course, a sharp-eyed theorist finds small key facts in natural phenomena and predicts that such and such particle should exist. A good example is the Yukawa meson; more recently, there was the prediction of a charm particle (a kind of quark) through the contributions of S. Glashow and several others. Still, it is not too much to say that nature always proves richer and more complex than the prediction. Though Yukawa believed that there is only one kind of meson there are actually a countless variety of mesons. The b quark was discovered despite the expectation by many physicists that charm should be the last of the varieties of quarks.

Because of such circumstances, particle physics always remains fresh and stimulating. Thus, also, the cost of constructing giant accelerators using national budgets may be justified. If all the fundamental particles were discovered, high energy physics as a world industry would probably end. But how can we say that we have already discovered all the particles? Our experience seems to show that we humans can never get to the bottom of the depth of nature all at once. Thus we must studiously continue our research.

The Principle of the Accelerator

Let me now explain what the accelerators are exactly. The fundamental principle of a particle experiment is quite simple; it consists of merely striking two particles against each other and seeing what happens. Generally various particles are created as a result of the collision. Among the particles created are unstable particles not usually seen in matter. Such particles "decay" into other particles soon after their creation. What they decay into and how long they take to decay (half-life) are important data along with the collision reaction itself.

The simplest example of an accelerator would be the picture tube in a television receiver. As everybody knows, a picture tube is that part of a television set that creates the images. As the cathode filament in the evacuated tube heats up, electrons fly out of the filament. By using high voltage (about 20,000 volts) these electrons are accelerated in a fixed direction towards the screen coated with fluorescent material. The electrons excite the atoms in the fluorescent material and a fluorescent reaction takes place; that is to say, photons (quanta of light) are emitted.

Although particles other than electrons such as protons (hydrogen nucleus) and other nuclei are accelerated in elementary particle experiments, accelerators, in general, can be separated into proton accelerators and electron accelerators. I spoke of volts when discussing television sets; it turns out, conveniently, that the same unit of measurement describes accelerators as well.

The unit eV (electron volt) is used to describe the energy of accelerators. For example, the cyclotron built at the University of Chicago by Enrico Fermi in the old days could achieve an energy of 4.5 MeV; MeV means million, or mega, electron volts. This is exactly the energy of an electron boiled off a cathode that has been accelerated through a potential of 4.5 million volts.

This energy is proportional to the charge e and the potential V and is therefore called eV. Since protons and electrons have the same electric charge except for the signs, the above applies to protons as well as electrons. In other words, cyclotrons apply the opposite polarity of the potential from television sets to accelerate protons. The energy needed to extract an electron from the outer shell of an atom, or the ionization energy, is, roughly speaking, of the order of a few eV's (The term "order" may be thought of as a rough way of characterizing a quantity by considering only the highest decimal places of a number). Therefore, the energy of a chemical reaction is in the order of a few eV's; it is no coincidence that the potential of a dry cell battery is 1.5 volts.

The nuclear binding energies, in contrast, are very large compared to atomic ionization energies and are of the order of millions of eV's. That is to say, a proton must have millions of electron volts of kinetic energy if it is to cause nuclear reactions when it strikes a nucleus. But it is technically impossible to apply millions of volts of potential at once, so particles are caught in the magnetic field of magnets to make them travel in circles, and they are gradually accelerated through a potential that is applied every time they make a circuit. It is exactly like kicking your legs at every swing when you are riding on a swing in a playground. This is the principle of the cyclotron invented by E. Lawrence. Even if only 1,000 volts are applied at each turn, the energy will rise to one million volts after 1,000 circuits. One could say that high energy experiments consist largely of catching particles that are produced in the collisions that occur between the target and the bullet particles when the latter are accelerated as above, and then released from the magnetic field in the direction of the target.

A device called a synchrotron, rather than the cyclotron, is used when the energy becomes larger than about 1 GeV (= 10^3 MeV). G means giga, or 10^9. The basic principle of this device is the same as that of a cyclotron; but a donut-shaped vacuum tube is surrounded by magnets and particles are accelerated inside the tube. The thickness of the donut may be a few centimeters but the circumference may be hundreds of meters. For example, the 12 GeV proton synchrotron at the High Energy Research Laboratory (KEK) in Tsukuba, Japan, has a radius of 17 meters and a circumference of 107 meters. The Fermi accelerator near Chicago reaches energies up to 800 GeV or more and has a radius of 1 kilometer, so it is as large as a track for car races.

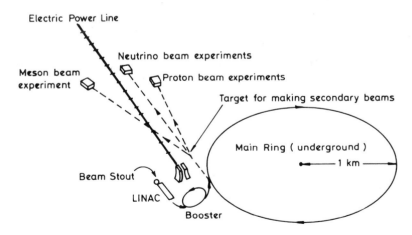

Fig. 4.1 An aerial view of the Fermi National Accelerator Laboratory, Batavia, Illinois. The largest circle is the main accelerator. Three experimental lines extend at a tangent from the accelerator. The 16-story twin-towered Wilson Hall is seen at the base of the experimental lines. (*Courtesy of the Fermi National Laboratory*)

To reach higher energies, proportionately larger accelerator radii are needed. This is because, first of all, it becomes more difficult to bend the paths of particles as they gain energy. Since the strength of magnets is limited, radii of accelerators must be bigger. Another reason is the energy loss due to radiation; charged particles, as they change direction, lose energy by emitting electromagnetic radiation, so the accelerator must supply an energy larger than the radiation loss at every circuit. This energy loss is proportional to the fourth power of the total energy measured in units of the mass of the particle and is inversely proportional to the radius of the accelerator; therefore, it severely limits the acceleration of a light particle such as the electron.

Thus, one of the ways to accelerate electrons is not to put them in circular orbits but to put them in a straight track and apply a potential repeatedly at set distances; so the energy gained by the electron is proportional to the length of the track. This is called a linear accelerator (LINAC) and probably the most famous example of it is the machine at the research center at Stanford, SLAC; SLAC requires 3 kilometers of track to accelerate electrons to 25 GeV. KEK also has a 2.5 GeV linear accelerator but the primary objective for this machine is not the electron beam itself but the secondary production of vast numbers of photons from the electron beam which will make the machine function as a "photon factory".

When an accelerator, be it a synchrotron or a LINAC, becomes as large as many kilometers, a practical limit of feasibility may be reached because of the land required and the cost. Fermi once said jokingly that he predicted that by the year 1984 (the title of the famous novel by George Orwell) the accelerator would be the size of the earth. It must be counted fortunate for Man that neither Fermi's prediction nor Orwell's pessimistic predictions show signs of coming true.

Colliding Beams — The Power of the Counterpunch

Still, it is not true that yet higher energies cannot be reached without making the accelerator unreasonably large; there is one way to get around the difficulties. This is the method of colliding beams, and, as its name indicates, two beams moving in opposite directions are created and made to collide. Thus, there is no distinction between bullet and target anymore and both have become bullets. An example of this type of accelerator is the machine called PETRA at the research center in Hamburg called DESY. The same basic principles as for synchrotrons are involved, but electrons and positrons are prepared and put into orbit in opposite directions. Conveniently, particles with opposite charges bend in opposite directions in the same magnetic field. If each particle has a kinetic energy of 15 GeV, a total of 30 GeV can change into reaction energy.

The above does not simply mean that the energy available in the reaction is doubled in comparison with the fixed target beam. In the case of the fixed target beam, the bullet's kinetic energy is not entirely converted into reaction energy; a part of the energy is used up as the (recoil) kinetic energy to bodily scatter the target. It is easy to understand that the loss to recoil energy is not much if the mass of the target is large compared to that of the bullet but it is large if the target is light. The situation gets worse and worse as the bullet approaches the speed of light; this is because the bullet gets heavier and heavier due to relativistic effects. Therefore, the useful energy is not measured by the energy E of the bullet but by the center-of-mass energy W. In fixed target accelerators, the center-of-mass energy is proportional to the square root of the energy of the bullet; thus, the center-of-mass energy is only doubled even if the energy of the bullet is quadrupled. On the other hand, if the center-of-mass of the colliding particles is stationary as in the case of colliding beams, then $W = 2E$; in other words, the center-of-mass energy is proportional to the bullet energy. For example, the center-of-mass energy available when colliding 450 GeV proton beams with a stationary hydrogen target is 30 GeV. So the amount of energy available for a reaction is the same as that of colliding electron beams of 15 GeV each.

Of course, the interactions between two electrons and between two protons are different so the experiments themselves have different relevances; but in the case of electron interactions, there is no other practical way to approach such a large energy. This is because the mass of the electron is only $\frac{1}{1800}$ that of the proton and so the electron takes up a large recoil energy when used as a fixed target. The weak point of the colliding beam technique is the low frequency of interactions. This may be understood if one thinks about the difference between shooting a target with a machine gun and letting the bullets collide in mid-air. In order to increase the efficiency, the beam must not only have high energy but become denser and more compact; but that process has a limit so the created beams are not used right away but stored. The beam goes around and around in the donut-shaped vacuum until enough particles are stored.

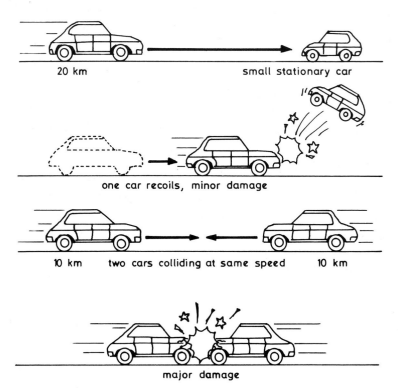

Colliding beams: if a large car strikes a small stationary car the damage is minor. If two cars going at the same speed collide head on there is major damage.

These are then the workings of accelerators as far as the main principles are concerned. Let's summarize:

Accelerators are classified by the method of acceleration into synchrotrons and LINAC's (linear accelerators), and by the experimental methods into fixed target types and colliding beam types. If classified by particle types, there are electron accelerators and proton accelerators; colliding beams therefore can be proton-proton, proton-antiproton, electron-electron, electron-positron (electron-antielectron), electron-proton, and so on, but the only kinds that actually exist are the proton-proton type, the largest of which is called ISR at the Joint European Research Center (CERN) in Geneva, and the proton-antiproton type, of which the only existing one is also at CERN and is called SPS. There is also an accelerator of this type under construction at Fermilab. As for the electron-positron type, the biggest is the PETRA at Hamburg; but CERN is planning a machine called LEP which will have energies in the 100 GeV's. The TRISTAN accelerator being built at KEK in Tsukuba is an electron-positron machine, at around 60 GeV.

Thus, the construction of particle accelerators, at a larger and larger scale, has become an international competition. Fermi's joke aside, the accelerator energy has increased ten-fold every 7 or 8 years for the past 50 years since Lawrence invented the cyclotron; now the energy is in the order of TeV's (10^{12} electron volts; T stands for tera). This progress has been made possible mainly through the improvements in technology. The actual cost of the construction has not drastically increased because of the constant influx of new methods. There is no predicting how long this fortunate situation will last, however.

5

THE BIRTH OF THE YUKAWA THEORY

From the Atom to the Nucleus

It can be safely said that the theory of the structure and characteristics of the atom was completed by the appearance of quantum mechanics. The problem of what the atoms are actually made of was answered earlier in 1911 by the brilliant inference of E. Rutherford. He concluded, from the character of scattering of alpha particles from atoms, that there is a heavy positively charged nucleus in the center of the atom, which is surrounded by clouds of electrons. But it is not possible, within the framework of classical electrodynamics, to understand exactly what the electrons are doing or why they do not gradually lose energy through radiation of electromagnetic waves.

In 1912, Niels Bohr introduced a quantum mechanical model of atoms and made a bold hypothesis that electrons only orbit at certain "quantized" orbits, and, when they radiate, they jump discontinuously from one orbit to another. This hypothesis in large part explained the nature of the spectrum of light emitted by atoms; but it was still a naive theory. The true understanding of the stability of the atom came in 1925 with the appearance of the quantum mechanics advocated by W. Heisenberg and E. Schroedinger.

Simply put, quantum mechanics states that the electron has a wavelike nature and forms something like a stationary wave around the nucleus. In terms of everyday phenomena, the wave pattern is like that of a piano or violin string which must vibrate at a characteristic frequency, which is either its fundamental frequency or one of its harmonics, i.e., some integer multiple of the fundamental frequency. Such is the case also for the electron wave in the atom, and its characteristic frequency is determined by Schroedinger's wave equation.

Quantum mechanics not only solved the mystery of the atom but also revolutionized the method of description of natural phenomena. Newton's classical laws of motion are replaced by the quantum mechanical laws of motion, but this is not simply a matter of exchanging one set of rules for another; the whole method of description is different. The velocities and positions are not directly determined by the laws of motion. Rather, it is indirectly determined by something called the probability amplitude which gives the probability distribution of the above quantities. But let me not go into this subject further since the purpose of this book in not to explain quantum mechanics. The reason I mention quantum mechanics is to show how, previously, physicists thought about its range of applicability.

Quantum mechanics solved the problems of atomic physics. That is to say, the behavior of atoms can be explained by applying quantum mechanics after assuming the existence of electrons and nuclei. But the nucleus itself was still not understood. The nucleus does not seem to be an elementary particle but is made of many protons and neutrons; the force that holds these particles together cannot be the electromagnetic force but some other force with completely different characteristics. Thus, the descriptive method of quantum mechanics might not work. The energy scale is different as well; the kinetic energy of the electrons in an atom is of the order of a few electron volts and the wavelength corresponding to this energy is about 10^{-8} cm, which is about the size of the atom. The nucleons in a nucleus differ in both energy and wavelength from atomic electrons by as much as a factor of one million. The physicists before Yukawa generally thought that the solution to the mysteries of the nuclei must involve all these circumstances and therefore must be quite difficult.

This thought is correct in a certain sense. Even now, it is impossible to completely describe nuclear forces beginning with a fundamental equation. But since we know that nucleons themselves are not elementary, this is like asking if one can exactly deduce the characteristics of a very complex molecule starting from Schroedinger's equation, a practically impossible task.

As I see it, the contribution of Yukawa lies in completely trusting the existing theoretical structure, that is to say, quantum mechanics and special relativity, and tackling the general problem of how to describe the short-range forces such as the nuclear force. The resulting answer calls for the existence of an unknown particle. But if this is a theoretically natural and compelling result, one should not fear even if there is no experimental evidence for it.

History proved Yukawa's prophecy to be correct. Let me here quote from Yukawa himself on his thoughts at that time.[a]

Looking back, I was strangely full of self-confidence in the autumn of 1934 when I arrived at the existence of mesons as the consequence of the theory of nuclear forces. Today, in theoretical physics, the Poincaré type of thinking which builds up from a hypothesis is the general rule. I do not think, however, that Descartes' method of starting from self-evident truths has become completely meaningless. Sometimes an idea that arises when one is continuously, even abnormally, concentrating on one problem begins to seem self-evident. Thus, one becomes self-confident, and desires to advance the idea a step further. Such ideas are, objectively considered, determined true or false only after their conclusions have been compared with the empirical facts. But for the man who thought of it, at least for some time, it is a "truth for which there is no other possibility". It is not possible to discuss creativity, especially the kind of creativity involved in discoveries in theoretical physics, with a person who considers only one side of the above duality.

Yukawa's Meson Theory

What, then, is Yukawa's meson theory? Let me now explain. The starting point for thinking about the nuclear force are the forces we understand already, namely gravity and electromagnetism. As stated before, these forces reach infinite distances, and their potential energies are inversely proportional to the distance. This situation arises, speaking mathematically, because both the gravitational and static electric fields obey Laplace's differential equation. But Laplace's equation only applies to cases where the source of the field is static. In more general cases, the fields are described by Maxwell's equations in the case of electromagnetism, and Einstein's equations in the case of gravity. One of the characteristics of these equations is that they admit waves that move with the speed of light.

In the late 19th century, James Clark Maxwell synthesized the various laws of electromagnetism discovered by the pioneers like Michael Faraday, made some mathematical corrections and succeeded in building a structure free from inconsistencies. As a

---

[a] Preface to Sosuke Takauchi, *Order and Chaos: The Thoughts of Hideki Yukawa*, 1974, Kosakusha edition.

result, the existence of the above-mentioned waves became inevitable. As the speed of these waves coincided with the speed of light, it was theorized that light must be an electromagnetic wave. This was eventually confirmed experimentally.

Maxwell's electromagnetic theory is the first of what are called field theories. Today, it is thought that all forces can be described by a quantity known as the field of force. This is, simply speaking, something that determines the force acting on an object placed at a certain point in space-time (certain point at a certain time); thus, the field of force is something that reflects the characteristics of a kind of medium called space-time.

In Newton's theory of gravity, on the other hand, it is thought that the gravitational force between two masses is a directly acting "action-at-a-distance". But this can be treated as a field theory as well; it can be interpreted that the first mass affects the surrounding space and creates a gravitational field, and the field applies the force to the second mass. The so-called gravitational potential is this field.

As shown by the above case, the action-at-a-distance point of view and field point of view are not necessarily contradictory;

HIDEKI YUKAWA (1907–1981)

Born in Tokyo, he grew up in Kyoto. His father was the geographer Takuji Ogawa. His was a family of scholars and he learned Chinese scholarship from his early years. While he was a student at Kyoto University, Heisenberg, Schroedinger, and others were in the process of formulating quantum mechanics; there were no teachers on the subject in Japan and he must have had to study it on his own. Perhaps from the spirit of independence nurtured in this period, Yukawa conceived of the meson theory in 1934 when he was 27, beating the European scientists to the draw. In 1949, he became the first Japanese to receive the Nobel Prize. As the leader of the Kyoto theoretical particle physics group he trained many scientists. After the war, he championed the non-local field theory as a way to overcome the deficiencies of the conventional field theory. His popular writings are also well known in Japan.

but since it has been found, as Maxwell's theory implies, that the electromagnetic force is "transmitted" at a finite speed, the concept of the field of force of a medium is more attractive. Maxwell's equation contains important keys in determining the properties of this medium. The most prominent key is the fact that the speed of light does not depend on the frame of reference. Albert Einstein was the one who spotted this key and claimed that this should apply to all physical phenomena, not just electromagnetism. Thus, according to Einstein's theory of special relativity, since the characteristics of the medium have been established by Maxwell's equations, all other fields must play by the same rules of the medium.

Einstein showed that the dynamics of a particle obey the special theory of relativity, but in quantum mechanics, particles also possess wave characteristics; the wave function that describes a particle can be thought of as a kind of field. The wave function obeys the Schroedinger equation, but since this equation is a quantum mechanical interpretation of Newton's classical mechanics, it is not within the framework of special relativity. Thus Schroedinger's equation represents non-relativistic quantum mechanics.

Relativistic Quantum Mechanics

What, then, will be the form of the relativistic quantum mechanics? What is the thing that replaces Schroedinger's equation? These questions were first answered by P.A.M. Dirac. The answer is the Dirac equation which determines the wave function of electrons.

The Dirac equation contains many new characteristics not contained in Schroedinger's equation. For example, Dirac electrons have a characteristic called spin, which can be thought of as expressing either something like the spin of a top or some internal state like the polarization of light. In any case, spin corresponds to the electron's intrinsic angular momentum and its magnitude can only be $\frac{1}{2}$ in units of Planck's constant, and its direction can only be up ($+\frac{1}{2}$) or down ($-\frac{1}{2}$). On the other hand, deciding which is up and which is down is arbitrary, just like resolving the polarization of light into two arbitrary perpendicular directions.

Although the spin of the electron was theorized prior to Dirac to explain phenomena like the atomic spectra, it was a great victory for relativistic quantum mechanics to have the spin come out automatically from the Dirac equation.

Another new result of the Dirac equation was the prediction of the existence of the "antielectron", or "positron". Thus, the positron state with charge opposite the electron's is also included in the Dirac wave function. If an electron (e^- or e) collides with a positron (e^+) they sometimes scatter but they can also annihilate and turn into several photons (γ, gamma); in equation form:

$$e^- + e^+ \rightarrow e^{-\,\prime} + e^{+\,\prime},$$
$$e^- + e^+ \rightarrow \gamma + \gamma^{\prime}.$$

P. A. M. DIRAC (1902-1984)

Born in England, he was one of the main contributors to the formulation of quantum mechanics. He was a genius whose work is unique in its mathematical beauty and depth. He was 26 when he discovered the Dirac equation which describes the electron relativistically. To explain the fact that this equation had negative energy solutions, he conceived of the interpretation that a negative energy solution represents an antielectron, and thereby predicted the existence of the positron. He received the Nobel Prize in 1933. He has since contributed enormously to the development of quantum theory, but one aspect of his work that became an important topic was his magnetic monopole theory which he advanced in 1931. By the Dirac equation he created the mathematics of spinors, and by the monopole theory he created the mathematics of fibre bundles, independently of the mathematicians. He said in his paper on the monopole that "it is inconceivable that nature will not utilize a mathematically beautiful and elegant theory." He recently passed away in Florida. His lectures were delivered without any notes and without wasting one word, and were as elegant as his papers.

Of course, these equations proceed also in the opposite direction. Thus, if there is enough energy, we can make an electron-positron pair. The positron was discovered in the U.S.A. by C. Anderson and its discovery confirmed the validity of the Dirac equation. It does seem peculiar, however, that we mostly see electrons in everyday phenomena and positrons only rarely. The usual answer to this is that a positron annihilates an electron as soon as it meets up with one; still, it is peculiar that there are so many more electrons than positrons in the world. The same thing can be said about protons and antiprotons. The attempt to answer this question will take us to the beginning of the Universe; but let us defer this discussion till later.

Let us proceed now to the discussion of the generalization of relativistic quantum mechanics. So far, I've explained that two of the characteristics of the Dirac equation are spin and the existence of the positron. But the Dirac equation is not the only relativistic equation; Maxwell's equations which describe the electromagnetic field are also relativistic. In quantum mechanics, even classical fields have particle characteristics, and classical particles also have field (wave) characteristics; thus both field and particle must be considered in a unified way. Such considerations lead to the following results.

The fundamental characteristics of particles such as the electron or the photon are mass, spin and electric charge; if these are determined, some appropriate wave equation can determine the motion of the particle. And each particle must, in general, have its antiparticle which has an opposite electric charge. Sometimes, however, there are cases when particle and antiparticle are one and the same. The quantum of the electromagnetic wave, the photon, is such a particle. In these cases, the electric charges must be, of course, zero. But it is important not to get confused about the statement that the photon's electric charge is zero and the statement that charged particles emit photons (electromagnetic radiation). The meaning of the first statement is that light itself cannot be the source of light.

The mass of the particle is, of course, unique to each kind of particle and can also be zero; if this is the case, the particle always travels at the speed of light. The photon is the primary example of such a particle; the quantum of gravity (the graviton), though not yet observed, should also be such a particle. Of the matter particles, the neutrino was thought to be a zero-mass particle but this has come under suspicion lately (Chapter 21).

Let us now discuss spin. As related earlier, the spin of the electron is $\frac{1}{2}$ (in units of Planck's constant) and its state can either be up $(+\frac{1}{2})$ or down $(-\frac{1}{2})$. In the case of the photon, the spin is 1. I said before that the spin is like the polarization of light, but in this case light polarization equals spin literally. Since we cannot stop light, we can not say the polarization is up or down; the angular momentum of the right or left-circularly polarized light with respect to its direction of motion is ± 1, and this is the analog of the two polarization states of the electron.

The spin, however, is not only $\frac{1}{2}$ or 1. In principle, spin can be 0, $\frac{1}{2}$, 1, $\frac{3}{2}$, . . . and so forth. In general, if the spin is j, then particles with mass has $2j + 1$ states and the magnitude of a component of the spin vector can take on the values $j, j - 1, j - 2, \ldots, -j$. For example, if the spin is 1, its component in the z-direction can be 1, 0, or -1. Massless particles such as photons can have only two states $\pm j$ (with respect to the direction of their propagation).

Now we must introduce Pauli's law of spin-statistics connection which divides the particles into two different groups: those with integer spin such as 0, 1, 2, . . . and those with half-integer spin such as $\frac{1}{2}$, $\frac{3}{2}$, . . . Simply stated, the half-integer spin particles such as the electron exist as particles also in classical mechanics and are the structural components of matter.

In contrast, the quantum of a field like the electromagnetic field which appears classically as a wave or a field of force, has integer spin. (The reverse statement is not true; this subject will be discussed later.)

The other distinction of the two groups of particles has to do with quantum statistics; half-integer spin particles obey Fermi statistics, and integer spin particles obey Bose statistics. Again speaking simply, particles that obey Fermi statistics cannot be in the same state at the same time; for example two of the same particles cannot occupy the same space with the same direction of spin. Thus, many particles cannot overlap in the same state, which fits in with our concept of "matter".

Particles that obey Bose statistics, however, have no limit to the number of particles that can be in the same state. This can be interpreted as implying that an unlimited number of waves of the same shape can overlap and make the overall amplitude of the wave arbitrarily large. Thus, although a single photon has a very small amount of energy, if many photons are together in the same state, they can be observed in everyday phenomena as electromagnetic waves. A good example of the above is the laser, which can be thought of as light composed of vast numbers of photons, all occupying exactly the same state.

Particles that obey Fermi statistics are called fermions (Fermi particles) and those that obey Bose statistics are called bosons (Bose particles). The former is named after the famous Enrico Fermi and the latter after the Indian physicist M.K. Bose.

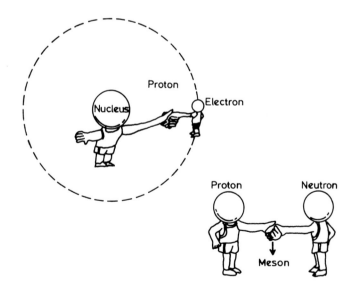

What holds together the proton and the neutron which are not electrically attracted to each other?

Other elementary particles such as the nucleons and their aggregates, i.e., the nuclei, also obey this general principle. For example, the nucleons, namely the neutron and the proton, and their antiparticles, or the antinucleons, are fermions of spin $\frac{1}{2}$ and are described by the Dirac equation. on the other hand, the heavy hydrogen nucleus which is composed of a neutron and a proton is a spin-1 boson; but in this case the particle is still a matter particle and not a force field. Thus it is not completely correct to characterize the difference between a boson and a fermion by saying that one of them comprises matter and the other is a field of force; but one might say that this is because the heavy hydrogen nucleus is not a truly elementary particle. By the same token, the hydrogen atom which is composed of an electron and a proton is also a boson.

The Ideas that Led to the Mesons

Are there, then, other elementary particles which are bosons? If there are, these must manifest themselves as fields of force. What about the nuclear force, for example? According to the general theory above, bosons may have masses as well; something called the Klein-Gordon wave equation describes a massive spin-zero boson and also expresses a short-range force field. The range of the force is inversely proportional to the mass of the boson and is equal to what is called the Compton wavelength of the particle.

The nuclear force, in fact, has the characteristics of a short-range force. Thus the nuclear force field is described by the Klein-Gordon equation and its quantum must exist as a massive boson. The range of the nuclear force is known roughly; it is about the size of a nucleus which is about 10^{-13} cm. The mass of the quantum of the nuclear force field can be calculated from the range, and this mass turns out to be about 200 times the mass of the electron (or $\frac{1}{10}$ of the mass of a nucleon); this was Yukawa's theory. Such a particle was unknown at that time, but Yukawa boldly predicted its existence and called it meson, meaning that it possesses a mass midway between the electron and nucleon masses.

The Yukawa theory was introduced in 1935. The birth of quantum mechanics was 10 years earlier in 1925. In the interval, our knowledge about the nuclei had advanced, and it was known that nuclei were made up of protons and neutrons. The positron had also been discovered. So the time was ripe for the introduction of the Yukawa theory; but why wasn't it thought of earlier?

One might say that it is easy to criticize after the fact. But it is instructive to investigate how other people were thinking at about the same time; let us imagine that we are in the first half of the 1930's.

The 1930's

It was a great success for quantum mechanics to be able to describe the atomic structure, that is to say, the motion of electrons inside the atom; but how far does the region effectively described by quantum mechanics extend? Here exists a yet smaller compound entity called atomic nucleus which seems to be an aggregate of nucleons, but the mechanism of binding is completely unknown. Perhaps it is necessary to have a new mechanics to solve the problem of the nucleus, just as it was necessary to formulate quantum mechanics to solve the problem of the atom. This is the first question.

Next, relativistic quantum mechanics makes possible the existence of many different kinds of particles, but the interpretation of antiparticles is not clear. The positron has been found, but what about the antinucleon? This world seems to contain only protons, neutrons, electrons (fermions), and the electromagnetic field and the gravitational field. In particular, particles with the mass of the meson have not been found. How far, in fact, can one trust relativistic quantum mechanics? This is the second question.

The greatness of Yukawa can be characterized, I think, by his ability to force the above issues. Simply put, he believed in the correctness and the reality of the two beautiful theories of the twentieth century, relativity and quantum mechanics, which had succeeded so brilliantly, and he applied the consequences of these theories directly to the atomic nucleus. It is not true that all physical phenomena are described by beautiful principles. It is also not true that nature always utilizes beautiful principles which are theoretically possible. But a beautiful and correct body of principles usually has wider application than initially expected; therefore, results that come out of such body of principles, even if seemingly contradictory to nature, must be taken seriously.

A good example of the above would be the black holes which are predicted from Einstein's general theory of relativity. I will not spend much time on black holes since it is out of the scope of this book; the story about the world in which time and space are warped to such an extent by the tremendous force of gravity that even light cannot escape is even more mysterious than the quark, and is of great interest to the public. The black hole recently became a popular topic because the advances in astronomy have made it possible to obtain several indirect pieces of evidence that black holes may exist in our Milky Way Galaxy. But originally it was Schwarzschild who found the state of a black hole as a solution to Einstein's equations for the gravitational field (general theory of relativity) shortly after the introduction of the theory by Einstein (1916).

6

THE APPEARANCE OF
NEW PARTICLES

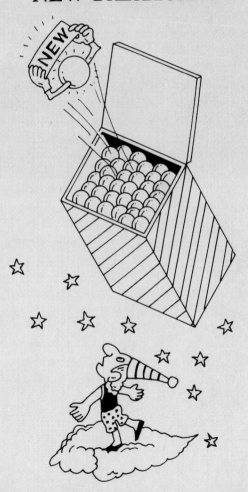

Coulomb Form and Yukawa Form

The story of the Yukawa theory that began in the last chapter is still not finished; only the fundamental concepts have been introduced. So far we have said that the nuclear force should be described by relativistic quantum mechanics, and, therefore, a boson called meson must exist. But there are many questions still to be answered. For example, why are mesons not found in the everyday world? This is one of those questions.

The following picture is often utilized when explaining the quantum mechanical influence of a force field on a particle; for example, the electromagnetic force (Coulomb force) between two electrons may be said to arise out of the exchange of the quantum of the electromagnetic field, the photon. Two electrons are playing "catch" with the photon as the ball. The fact that the mass of the photon is zero (thus moving with the speed of light) corresponds to the long-range Coulomb form of the force, and the potential energy is inversely proportional ($1/R$) to the distance R.

But if a massive boson, a heavy ball, was used in this game of catch, the force cannot reach very far. In this case, the poten-

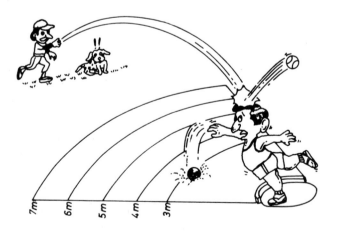

A heavy ball cannot be thrown far.

tial of the "Yukawa" form. Written in equation form, it is given by the expression

$$V \propto \frac{e^{-\mu R}}{R}$$

which means that the force weakens exponentially far away. Here μ is a quantity which is proportional to the mass of the particle, and $1/\mu$ is the Compton wavelength. It can be said that the range of the force is equal to the Compton wavelength. Unfortunately, how such a form of the potential comes about cannot be explained without the knowledge of differential equations; it is possible, however, to make the form somewhat plausible by the use of Heisenberg's uncertainty principle.

The particle being exchanged here is not a "real" particle; that is to say, there is not enough energy in the initial state to let the boson have mass. Such particles are called "virtual" particles and they only exist temporarily during this game of catch and fade away again quickly. According to the uncertainty principle, the uncertainty $\triangle T$ in time and uncertainty $\triangle E$ in energy are inversely proportional to each other:

$$\triangle E \times \triangle T \gtrsim h \text{ (Planck's constant).} \qquad \text{(a)}$$

Thus if the lifetime $\triangle T$ is small, the uncertainty in energy becomes large and virtual particles can be made. But that particle can only travel at most a distance $\triangle T \times c$ (c is the speed of light) in the time $\triangle T$. If $\triangle E$ is the rest mass of the particle (mc^2), the range of the particle is $h/mc = 1/\mu$; this is the Compton wavelength.

The foregoing argument roughly explains the range of the Yukawa type force; the sign of the force, i.e., whether the force is attractive or repulsive depends on the sign of the "charge" of the source of the force. As the sign of the electric charge is proportional to the Coulomb field in the case of the electromagnetic field, the source of the Yukawa type force must also have a unique constant. In general, such a constant is called a coupling constant and its value determines the magnitude of the force.

The reason the nuclear force is much stronger than the electromagnetic force is that the coupling constant is much larger than that for the electromagnetic force. The coupling constant for the electromagnetic field is the unit charge e; changing this to a dimensionless number, the approximate strength of the force is given by the famous fine structure constant $e^2/\hbar c = \frac{1}{137}$.

The fact that this number is much less than one is thought to indicate that the electromagnetic force is a relatively weak force.

Since it is known that the nuclear force is much stronger than the electromagnetic force, the coupling constant of the meson ought to be larger than $e^2/\hbar c$. Although it is known that the coupling constant for the meson is in the order of $\frac{1}{10}$ to 1, another factor, namely the spin of the boson, also influences the nature of the force. For example, the force that arises from the exchange of a spin-1 neutral particle, such as the photon, is repulsive between two particles of like sign, but the force due to the exchange of a spin-0 particle is attractive. (Why this is so is a little hard to explain and will not be discussed.)

According to Yukawa's hypothesis, the spin of the meson is 0 and the charge is ± 1. If "catch" is played with a ball carrying charge, the charge of the two particles must change. If a proton throws a ± 1 charge meson at a neutron, the proton becomes a neutron and the neutron a proton. Thus, if such an exchange is allowed, the proton and neutron lose their identity and become merely two different states of the nucleon. Also in this case, there can be no game of catch between two protons or two neutrons.

In any case, the fact that the distinction between the proton and the neutron has disappeared has important implications. This following type of thinking began with Heisenberg. As already stated in Chapter 3, the proton and the neutron are very much alike except for the charge; let us ignore small differences when talking about the nuclear force and assume that their masses are equal. But then some identifying mark (a quantum number) is necessary to distinguish one from the other. In analogy with the spin which has two directions, up and down, let us introduce something called isospin, and let its up state and down state correspond to the proton and neutron. (The word isospin is a combination of isotope, meaning "same form", and spin. But atomic isotopes refer to atomic nuclei with the same atomic number but different mass. This definition of isospin is not the same as above.) Thus the isospin of the nucleon is $\frac{1}{2}$ and its z-component, depending on the direction of the isospin, is $+\frac{1}{2}$ or $-\frac{1}{2}$. This is denoted $I = \frac{1}{2}, I_z = \pm \frac{1}{2}$.

According to our present knowledge, the nuclear force is not simply due to one kind of meson, but it is a complex process involving the exchange of many kinds of mesons of various masses, spins and charges. But the lightest meson, having the longest range, is thought to be the most important. This meson is the pi (π) meson or pion, and was discovered first; it is generally thought that the discovery of the pion confirmed the Yukawa theory.

The π-Meson

The π-meson has a mass 270 times that of an electron, is spin 0 and, contrary to the initial expectation of Yukawa, has parity minus. Parity is a characteristic concerning reflection in space; when "looking into a mirror" or changing from a right-handed coordinate system to a left-handed one, parity is said to be plus if the field (potential) changes sign, if not, it is minus.

Also, π-mesons may have three kinds of electric charges: ± 1 and 0. Thus there are cases when nucleons exchange charge and also cases when they do not; as a result, the nuclear force has a characteristic of not being dependent on the direction of the isospin, i.e. whether it is between two protons, two neutrons, or a proton and a neutron. The three pions (π^0, π^+, π^-) may be said to have isospin; the value of the isospin is $I = 1$, $I_z = 1, 0, -1$. These π-mesons are produced as "real" particles when particles collide at high enough energies. For example:

$$p + p \rightarrow p + n + \pi^+,$$
$$p + p \rightarrow p + p + \pi^0. \tag{b}$$

π-mesons have a mass of about 140 MeV in energy units so, taking the energy lost to the recoil of the target nucleon into account, a cyclotron of energy above 270 MeV is needed for the above reactions. But π-mesons made thus are not stable particles. $\pi^{\pm}$ will decay into a muon (μ) and a mu-neutrino (ν_μ); π^0 will decay into two photons (γ).

$$\begin{aligned}
\pi^+ &\to \mu^+ + \nu_\mu \\
\pi^- &\to \mu^- + \bar{\nu}_\mu \\
\pi^0 &\to \gamma + \gamma
\end{aligned}$$ (c)

($\bar{\nu}_\mu$ is the antineutrino, the antiparticle of ν_μ, the neutrino.) The lifetime of the charged π-meson is 10^{-8} sec.; that of the neutral π-meson is 10^{-16} sec. These short lifetimes are the reason why Yukawa mesons are not floating around in the world.

Yukawa theorized that the meson decays into an electron and a neutrino, and tried to explain the neutron beta decay. Thus:

$$n \to p + \text{``}\pi^-\text{''}, \qquad \text{``}\pi^-\text{''} \to e^- + \bar{\nu} .$$ (d)

Here the "π" is the hypothesized π-meson in its virtual state and corresponds to an intermediate stage of the decay. But in reality, the π-meson decays first into a muon (μ) and a mu-neutrino (ν_μ), and the muon then undergoes a beta decay.

$$\begin{aligned}
\pi^+ &\to \mu^+ + \nu_\mu, \\
\mu^+ &\to e^+ + \nu_e + \bar{\nu}_\mu, \\
\pi^- &\to \mu^- + \bar{\nu}_\mu, \\
\mu^- &\to e^- + \bar{\nu}_e + \nu_\mu.
\end{aligned}$$ (e)

Here the ν_μ and ν_e are the neutrino partners of the muon and the electron respectively, and are not the same particle.

The muon was discovered in 1939, after the introduction of the Yukawa theory, by C. Anderson (who also had discovered the positron) and others, in the cosmic rays and was promptly interpreted as Yukawa's meson, the π-meson. This is understandable: there is no great difference in mass between the two (the muon's mass is $\frac{3}{4}$ of that of the π-meson). Also, the Yukawa meson is necessary to explain the nuclear force but the muon has no apparent purpose in nature.

The muon was discovered because cosmic rays that come from outside the earth collide with the nuclei in the air, producing π-mesons which decay immediately to muons. The reason muons do not likewise decay right away is because their lifetime is 100 times that of the pion (the time also becomes longer by the principle of relativity if the particle is travelling fast) and also because the muon is a lepton, like the electron, and thus does not interact strongly with the matter in the atmosphere.

Without knowing the above two-step process, we run into contradictions in interpreting the phenomena. The strong interaction is necessary to make enough pions in the the upper atmosphere. The representative reaction of this kind is:

$$p_1 + p_2(n_2) \rightarrow n_1 + \pi^+ + p_2(n_2). \tag{f}$$

This indicates the proton p_1 colliding with the proton p_2 (or neutron n_2) in nuclei in the air, and, as a result, changing into the neutron n_1 and creating the π^+-meson. But if this is true, then the opposite reaction, i.e., π^+-meson absorption by a neutron in a nucleus ($\pi^+ + n \rightarrow p$), should have the same rate, the result being that pions do not easily reach the earth's surface.

From Two-Meson Theory to the Discovery of the Muon

This mystery puzzled physicists for several years; the right answer was suggested by Y. Tanikawa and S. Sakata in a theory called the two-meson theory. In fact the version proposed by Sakata and T. Inoue had been almost on target. But the two-meson theory, having been developed during the war, did not interest people for a long time. The pion was actually discovered after the war in 1947. A group in England led by C.F. Powell succeeded in developing a highly sensitive photographic plate, which was exposed to cosmic rays. Examining the developed emulsion under a microscope, they found the tracks left by the cosmic ray particles in the emulsion. Thus they captured on film the following chain of reactions:

$$p + Z \rightarrow \pi, \qquad \pi \rightarrow \mu, \qquad \pi \rightarrow e.$$

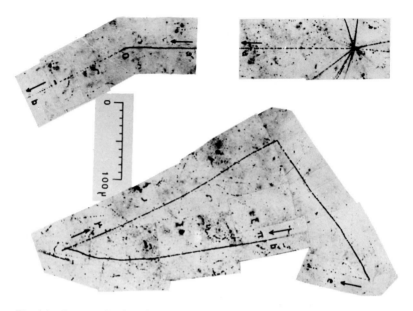

Fig. 6.1   An example of particle tracks in emulsion. (*Courtesy of Prof. R. Levi-Setti*)

(Z means an atomic nucleus.) A relatively low energy pion was created and, luckily, ran parallel in the emulsion layer while gradually losing energy and eventually coming to rest; then the pion decayed into a muon and it too, after a similar process, decayed into an electron and two neutrinos. (Neutrinos, of course, do not leave tracks.)

In general, the thickness of the track is determined by the ability of the particle to ionize the atoms; this ability weakens as the speed of the particle increases and reaches a minimum when the particle approaches the speed of light. It was because of the advance in technology that it was possible to record such weakly ionizing tracks that made the discovery of the Yukawa meson possible.

The discovery of the pion and the muon had an enormous symbolic impact in that they changed people's ideas about elementary particles completely. Up to that time, the elementary particles were the fundamental constituents of the objects in our everyday world and were stable. (It is strange to call something unstable an elementary particle.) But there were some exceptions; neutrons are stable in nuclei but undergo beta decay into a proton if left by themselves. At that time the neutrino was not quite out of the realm of conjecture, but it could not be called a constituent of matter. The fact that other elements like radium also give off beta radiation seemed to suggest that nature had some principles we did not understand.

But with the discovery of the muon and the pion, the exceptions became exceptions no longer. Both of these new particles are unstable and are not found in normal matter. The muon is like a heavy electron and belongs, in modern language, to the lepton family; but why does it exist? This was quite mysterious especially since the only leptons known, until quite recently, were the electron, the muon, the electron neutrino, and the muon neutrino.

But the existence of pions showed that the strong force called the nuclear force is not something mysterious but fits within the framework of relativistic quantum mechanics. Is the pion the only particle that mediates the nuclear force? If it is "natural" to assume that nature ought to be simple, then, of course, the pion should be the only one; under this assumption, the physicists of the time made a lot of effort in trying to explain the nuclear force in detail.

In reality, not only the nuclear force could not be fully explained, but also the pion turned out to be merely the first one of the family of mesons, and many more mysterious mesons and baryons began to be discovered one after the other. Thus, the nuclei we know of are hadrons that just happen to be stable.

The Dramatic Appearance of the V Particle

Not long after the discovery of the π-meson, something called the V particle was discovered; again in cosmic rays. This discovery was not made in the photographic emulsion but in the old-fashioned cloud chamber. The particle left a track that looked like the letter V, hence the name. The same kind of fork in the track is observed when the pion decays into a muon, but the V particle had larger mass, and, also, there seemed to be more than one kind of them.

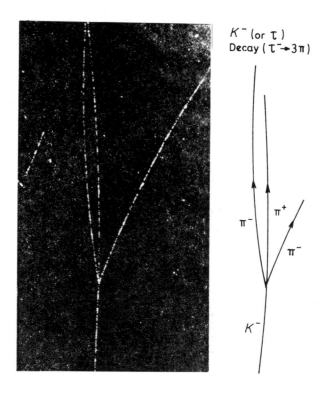

Fig. 6.2 V particle tracks. (*Courtesy of Prof. R. Levi-Setti*)

Some charged $V^{\pm}$ particles decay into a charged particle and a neutral particle; some neutral V^0's decay into two charged particles and leave a reverse V track. Other $V^{\pm}$ particles decay into three charged particles. The decay end products seem to be pions, muons, protons and so forth. Another peculiar thing is that there seem to be a lot of photographs with two V particles. How can such phenomena be interpreted?

This puzzle, I think, is a typical one in physics. I was myself at Osaka City University at the time and remember thinking about it quite a lot with my colleagues. The reactions in which a V particle is seen are quite rare among reactions of cosmic rays. The probability of seeing two V particles must be almost nil, if the two particles were independently produced. Thus the fact that two V particles are often seen in a pair must be thought to indicate that they were created by the same reaction. In fact, it must be that there is a necessity to make two such particles at the same time. Otherwise, the probability of producing one V particle must be vastly larger than producing two. Then why must these be produced in pairs?

Here one remembers the conservation laws we spoke of in the first chapter. We conjecture a conserved quantity from the regularity of the reaction. For example, the fact that an electron (e^-)-positron (e^+) pair can be produced but a pair e^-e^- or e^+e^+ cannot, is due to the conservation of electric charge. So the V particle, too, may carry some new kind of conserved quantity (quantum number).

This new conserved quantity was named "strangeness" by Murray Gell-Mann and the name has stuck. The name indicates that V particles are very rare.

The cosmic ray protons collide with matter and produce a positive strangeness-negative strangeness pair from a zero strangeness state. But according to experiments, one of the two V particles may decay into a proton; this must mean that there is a fermion V particle which is heavier than a proton and a meson-like V particle which is lighter. The fact that a pair is not always found must mean that one of them may leave the chamber without decaying or is missed for some other reason; this sort of situation occurs quite commonly in the laboratory.

But then what happens to the V particle once it is made? If strangeness is conserved, this quantity is conserved even if the particle decays into other particles and can never disapper. Other ordinary matter do not carry strangeness. There would be no contradiction if, for example, a strange neutral particle is produced, without leaving a track together with a nonstrange pion pair, but an analysis of the $\pi^{\pm}$ pair energies and angles showed that the decay was a two-body process.

Trying Everything

Now we are stuck again. One way out of the difficulty is to break the conservation law we just established. We assume that the conservation law is not completely strict but that there are some exceptions. This sounds quite opportunistic, something that should not be done by physicists who value exact theories; but if we use the term selection rule instead of conservation law there are many other examples like this. Nevertheless, since we started out thinking that strangeness is something like the electric charge, discarding the absolute conservation of strangeness makes us unhappy. But researchers must try everything. Moreover, once we start using this assumption, we immediately find various benefits.

First of all, since V particle decays are "almost" forbidden, the lifetime of a V particle should be relatively long; here relatively means in comparison with the time expected according to the uncertainty principle. If we substitute into the aforementioned uncertainty relation $\Delta E \times \Delta T \gtrsim h$, the energy that is released as the kinetic energy of the secondary particles of the decay (this is usually called Q), we arrive at a value of about 10^{-23} seconds for ΔT. In other words, if the particle moves at the speed of light during this interval, it can only travel 10^{-13} cm; this is about the size of a nucleus. It is a natural time scale when there is no conservation law, and it can be said that a particle with such a time scale is a "virtual" particle just like the particle involved in the game of catch in the nuclear forces.

The actual lifetimes of the V particles are quite a bit longer. Since the particles are created in the wall of the cloud chamber and decay after leaving tracks of several centimeters, their lifetimes must be at least about 10^{-10} seconds. This value is closer in order of magnitude to the lifetime of the pion which is 10^{-8} sec and of the muon which is 10^{-6} sec.

All of this reminds us of the fact that the decay of the muon and the pion are considered as a kind of beta decay. In particular, the muon decay is like the beta decay of the neutron because the decay products are rather similar. In fact, if what is called Fermi's theory of beta decay is applied to the muon, the lifetime comes out correctly. (The difference in the respective lifetimes of the neutron, 10 min, and of the muon, 10^{-6} sec, is due to the difference in the Q value; the basic parameter (coupling constant) in the theory remains the same.) Perhaps the decay of all the V particles may be considered a kind of beta decay. The reason the lifetime of the V particles is even smaller than that of the muon is because its Q value is even larger than that of the muon.

So there seems to be quite a variety to these rather odd phenomena called beta decay. In any case, these decays are thought to be slow reactions which imply small coupling constants. They also possess a rather bad-tempered habit of not giving importance to conservation laws.

Let me summarize what we have learned from nature.

1) There are many varieties of particles aside from those that make up the everyday objects.

2) There are also different kinds of forces in nature, which include the electromagnetic force and the nuclear (strong) force, and there exist particles that mediate these forces.

3) The weak interaction does not always respect conservation laws.

Regarding the last point, we spoke already about the non-conservation of strangeness; we have since found, as our knowledge increased, that other quantities are also not conserved in weak interaction. The most famous of them is the non-conservation of parity; we will take up this subject, along with the non-conservation of strangeness, in the next chapter.

7

ORDERLINESS OF ELEMENTARY PARTICLES AND CONSERVATION LAWS

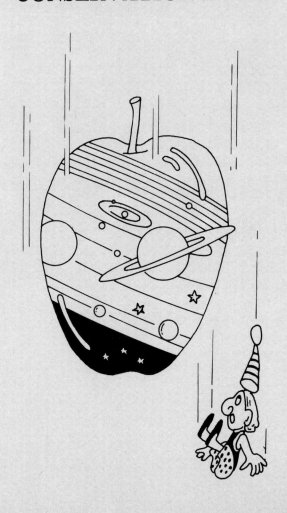

The Gell-Mann-Nakano-Nishijima Law

During the 1950's, after the discovery of the π-meson, the muon and the V particle, many new particles were discovered one after the other; the number of known particles increased explosively. But all of these new particles, with the exception of the muon, were hadrons; this trend continues to this day. The tau (τ), the next lepton after the muon, was discovered only a short time ago. It is rather convenient for us that there are a large number of hadrons. We can systematically examine the characteristics of the hadrons and look for some governing laws. For example, in the case of the atomic structure, the vast amounts of data available on the spectrum of light emitted by the atoms became an invaluable aid in the birth of quantum mechanics. Going back farther, it could be said that, because there were several planets in our solar system. Copernicus thought of his model of the solar system, Kepler discovered his laws based on the observational data of Tycho Brahe, all of which eventually led to Newton's laws of mechanics. There are other examples such as Mendeleev's periodic table and the nuclear shell model of Mayer, Jensen and Seuss.

In the case of the hadrons, the regularity that became the breakthrough is called the Gell-Mann-Nakano-Nishijima law (equation). To explain this law in a natural way, the hypothetical particle named quark was introduced. But let us start with the basic facts.

V particles, π-mesons and nucleons are among the "stable" particles of the hadron family. "Stable" means that the particles would not decay if weak and electromagnetic interactions were ignored. (There are not many cases of electromagnetic decay; two such examples would be π^0-meson decay and Σ^0 particle decay.) Thus these particles have long lifetimes, and hence we have a good chance of capturing their decay in emulsions or cloud chambers. The majority of hadrons are "unstable", meaning they decay by strong interaction, and they decay swiftly into a small number of "stable" hadrons; they are, so to speak, temporary excited states.

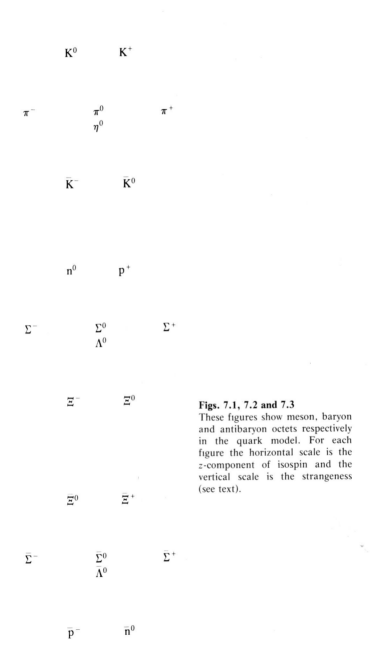

Figs. 7.1, 7.2 and 7.3
These figures show meson, baryon and antibaryon octets respectively in the quark model. For each figure the horizontal scale is the z-component of isospin and the vertical scale is the strangeness (see text).

"Stable" hadrons, as before, are divided into baryons and mesons, and the members of each group are shown on the previous page (Figs. 7.1–7.3).

These figures are organized according to the Gell-Mann-Nakano-Nishijima law. The first thing you notice about these figures is that each group of mesons, baryons, and antibaryons forms a group of eight. This is why Gell-Mann called this the Eight-fold way after the term in Buddhism. Here, the horizontal scale is the z-component of the isospin I_z; the vertical scale is the strangeness S, or what is called the hypercharge Y. (In the case of mesons, $Y = S$; in the case of baryons and antibaryons, $Y = S + 1$.) The electric charge (indicated by the superscripts), as can be seen, increases diagonally upward to the right.

The particles that line up horizontally are a family of particles with the same isospin I. These family members, as explained before for nucleons and π-mesons, are alike both in their strong interactions and masses. Looking at things more roughly, we can almost say that all eight of the particles in a group are the same. The masses are of the same order and their interactions are also similar. The differences associated with the difference in I_z is much less severe than those associated with the difference in strangeness.

The rough point of view such as that above is of great importance in physics; of course it is also a very human way of looking at things. This kind of thinking helps in finding hidden orderliness, and it also can become the foundation on which a very precise theory can be built. What then is the orderliness in our present case?

The thing that can be seen from Figs. 7.1 and 7.2 is that there is a relationship between the isospin I and the hypercharge Y (or strangeness). Let us take the case of mesons for example. The K-meson has $I = \frac{1}{2}$, $Y = 1$; the π-meson has $I = 1$, $Y = 0$; the η^0 has $I = 0$, $Y = 0$. Thus, if Y changes by 1, I changes by $\pm \frac{1}{2}$. Fig. 7.4 shows this.

Looking at the electric charge, it increases with regularity, as pointed out before, to the upper right; thus, the electric charge Q can be expressed simply in terms of I_z and Y

$$Q = I_z + Y/2.$$

This is what is called the Gell-Mann-Nakano-Nishijima (GNN) equation.

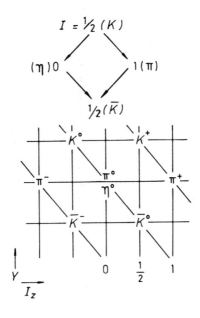

Fig. 7.4 A change of Y by 1 changes I by $\pm\frac{1}{2}$.

This may look quite simple after the fact, but the work and joy of physicists is in finding the correct pattern before knowing how many more particles there are. Fig. 7.5 summarizes the experimental facts known at the time of the appearance of the Gell-Mann-Nakano-Nishijima law.

Here the arrows indicate decay, and the particles produced in decay are indicated on the side. V particle masses were known roughly and the different positions of the V particles indicate that they come with different masses. Consider now the diagrams of Figs. 7.1 and 7.2 from 7.5. What do the V particles correspond to in the previous figures? I invite the reader to think about it. (The answer is Fig. 7.6.)

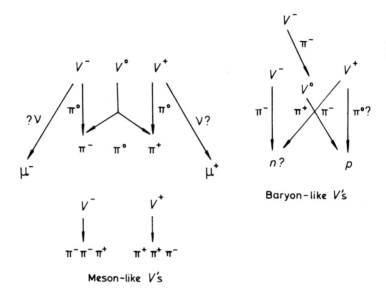

Fig. 7.5   Experimental facts known at the time of the appearance of the GNN equation.

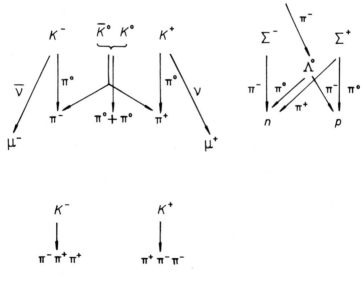

Fig. 7.6

The Key to an Idea

The equation found independently by Kazuhiko Nishijima and Tadao Nakano, and Murray Gell-Mann in 1953 cannot be inferred by simply looking at the preceding diagrams. The real key, as pointed out in the last chapter, is the idea that there is a new quantum number (hypercharge) which governs the production of the V particles but is violated in the decay, and one must add one other crucial component to this idea. It is the relationship between isospin and strangeness explained on p. 69, that a change of hypercharge by one is accompanied by a change of isospin I_z by $\pm \frac{1}{2}$. Assuming this, we notice that when a strange particle (particle carrying strangeness) decays into a normal particle by the emission of a pion, the conservation of isospin is violated as well as that of strangeness. Since the isospin of the pion is 1, the difference in the total isospin before and after the decay is a half-integer and cannot be zero.

Of course, we can say that there is no need to insist on breaking two conservation laws; but looking at Figs. 7.1 and 7.2, we notice that spin and isospin of a particle are simultaneously integers or half-integers; for example, the nucleon's (spin $\frac{1}{2}$) I is $\frac{1}{2}$, the pion's (spin 0) is 1. The V particles, on the other

---

KAZUHIKO NISHIJIMA

Born 1926. Toward the end of the war, S. Tomonaga was instrumental in starting a new particle physics theory group at the University of Tokyo; Nishijima was among the members of this group. He began to work at the newly founded Osaka City University when the data on strangeness was beginning to accumulate. The Nakano-Nishijima law came out in collaboration with a colleague Tadao Nakano; Murray Gell-Mann, in the U.S.A., came to the same conclusion independently. In the following years, Nishijima spent time in Germany and the U.S.A. and he is now a professor at the University of Tokyo. His other contributions include the two-neutrino theory (electron and muon neutrinos) and various theorems of field theory. He is also well-known as the author of various textbooks on particle theory and field theory.

hand, do not seem to have the same kind of correlation, at least at first glance. But perhaps there is a correlation; how about looking for an unconventional inverse correlation for the V particle? Then we must put in a few unknown particles in our diagram.

As a result of inferences such as above, the GNN equation was born, and the existence of new particles was predicted. After the predictions were fulfilled, the GNN hypothesis became established as a theory and physicists could go on to the next step. And everybody lived happily ever after, so to speak.

What then was the next step? The GNN equation relates different isospins but does not explain why there are eight in a group. Let me defer this problem to later chapters and first explain that the confirmation of the GNN theory involves more than the confirmation of the existence of the eight particles. The GNN theory includes an assumption that the strong interaction conserves isospin and hypercharge. Since the creation of the V particle is predominantly by strong interaction and other interactions are only involved in its decay, the conservation laws above should be checked in the creation of the V particle. I spoke of the conservation of hypercharge in connection with the explanation of pair production, but the conservation of isospin must be treated in the same way.

Strong Interactions and the Conservation of Isospin

Let us take, as an example, the πN reaction. This means hitting a proton target with a $\pi^{\pm}$-meson beam. According to the GNN equation (p. 69), isospin is related to charge and strangeness so the former is conserved if the latter two are conserved; in reality, the conservation of isospin is a stronger requirement than that; the probability of the reaction (scattering cross section) does not depend on the direction (component) of isospin.

For example, the isospin I of a nucleon is $\frac{1}{2}$ and the I_z of proton and neutron is $\frac{1}{2}$ and $-\frac{1}{2}$ respectively; the isospin of the pion is 1 and the I_z of π^+, π^0, π^- are respectively 1, 0, -1. So all possible combinations of the I_z's of the pion and the nucleon are $\frac{3}{2}$, $\frac{1}{2}$, $-\frac{1}{2}$, $-\frac{3}{2}$. Thus

$$I_z = \tfrac{3}{2} \qquad \pi^+ + p \rightarrow \pi^+ + p, \qquad\qquad \text{(a)}$$

$$I_z = \tfrac{1}{2} \qquad \pi^+ + n \rightarrow \pi^+ + n, \qquad\qquad \text{(b)}$$

$$\rightarrow \pi^0 + p, \qquad\qquad \text{(c)}$$

$$I_z = \tfrac{1}{2} \qquad \pi^- + p \rightarrow \pi^0 + n, \qquad\qquad \text{(d)}$$

$$\rightarrow \pi^- + p, \qquad\qquad \text{(e)}$$

$$I_z = \tfrac{3}{2} \qquad \pi^- + n \rightarrow \pi^- + n. \qquad\qquad \text{(f)}$$

Dividing these equations into an upper two and a lower two and exchanging the two parts upside down is exactly the same as flipping the direction of the isospin. From the conservation of isospin, the reactions which correspond to each other in the above operation should have the same scattering cross sections.

There is actually an even stronger condition imposed. I will not speak of details because it is difficult to explain without the use of mathematics; however, thinking of isospin as a vector, adding an $I = \tfrac{1}{2}$ vector (nucleon) and an $I = 1$ vector (π-meson) has only two quantum mechanical answers for the resultant I, which are $I = 1 \pm \tfrac{1}{2}$, or, $\tfrac{3}{2}$ and $\tfrac{1}{2}$. For example, the above reactions under $I_z = \pm \tfrac{3}{2}$ belong to the $I = \tfrac{3}{2}$ case and the reaction under $I_z = \pm \tfrac{1}{2}$ are cases which are mixtures of $I = \tfrac{3}{2}$ and $I = \tfrac{1}{2}$ states. Thus, the cross section can be expressed as the combination of the numbers related to each of the cases when $I = \tfrac{3}{2}$ and $I = \tfrac{1}{2}$. But, there are certain energies when one of the I values gives overwhelmingly large cross sections. This happens when there is a resonance in a particular I. It means the same kind of phenomenon as the resonance we talk about in, say, stringed musical instruments; if two particles collide at a certain resonance frequency (energy), they can be thought to resonate vigorously with each other, thus increasing the cross section for the reaction.

Let us say that there is an $I = \tfrac{3}{2}$ resonance in the πN (pion-nucleon) collision. This means that the six reactions above can be expressed solely in terms of the quantity associated with the $I = \frac{3}{2}$, and the ratios of all the cross sections are uniquely determined:

$a : b : c : d : e : f = 9 : 1 : 2 : 2 : 1 : 9$

This historically famous 3-3 resonance is the first to be discovered of the resonances in hadron reactions. (3-3 indicates an isospin $\frac{3}{2}$, spin $\frac{3}{2}$ state.)

The 3-3 Resonance of the Pion and the Nucleon

The new cyclotron at the University of Chicago began to work in the early 1950's. The planner of the project was Enrico Fermi; the cyclotron energy was 450 MeV. Its purpose was not only to study simple pp collisions but also to create π-mesons in order to study their characteristics. By putting the proton beam on an appropriate target, π-mesons are created by the reaction of the beam proton with the proton (p) or the neutron (n) in the target nucleus.

$p + p \rightarrow p + n + \pi^{+}$
$p + n \rightarrow n + n + \pi^{+}$

The kinetic energy of the π-meson is distributed from zero to 200 MeV but it can be resolved by using magnetic fields to make a secondary pion beam of selected energy. Since π-mesons will travel at least a few meters, before decaying, the plan of Fermi's group was to put the secondary pion beam on a target and carry out the πN scattering experiment.

A peculiar thing was discovered during this experiment. The scattering cross section increased as the pion beam energy was raised. Thereupon a young theorist, K. Brueckner, pointed out to Fermi that there may exist a resonance of the pion and the nucleon. The same thing was also pointed out by Y. Fujimoto and H. Miyazawa.

A resonance or resonant state can be thought of as a very short-lived particle. It carries a set of quantum numbers (spin, isospin, strangeness, etc.) like normal particles. A difference with regular particles is that its mass is smeared and has a width; this is based on the uncertainty principle between time and energy which implies that the shorter the lifetime, the wider the width in mass. In a strict sense, particles such as neutrons and pions can also be called resonances, but the resonances we talk about here have lifetimes far shorter than the above particles and will decay even as they travel the distance across a nucleus. Thus the widths themselves are of the order of a million electron volts which is comparable to the mass of the resonance.

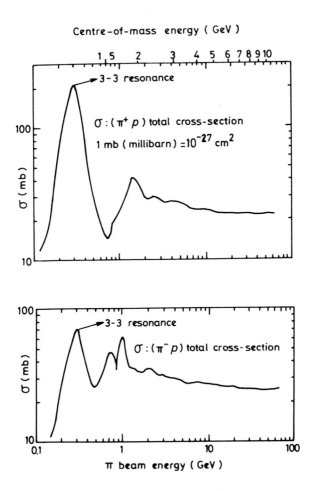

Fig. 7.7 πN resonance scattering.

The reason Fermi's πN scattering experiment indicated a scattering cross section rising with the energy was that the experiment had approached the energy where such a resonance began. Thus, there was an excited state of the nucleon, N*, which contributed to the following reaction:

$$\pi + N \rightarrow N^* \rightarrow \pi + N.$$

The typical behavior of a resonance is indicated in Fig. 7.7. The position of the peak of the mountain in that figure corresponds to the mass of the resonance (its median) and the width of the mountain corresponds to the width of the resonance. (The narrower the width, the taller the peak, and the resonance becomes more noticeable.) It can usually be concluded that a narrow peak in the scattering cross section corresponds to a resonance.

Since resonances carry set quantum numbers, it is important to determine the values of these numbers; A good method for determing the isospin is to compare the ratios of various reactions as we did in p. 74. In the case of Fermi's experiment the reaction ratio $a : b : c = 9 : 1 : 2$ determined that the resonant state had an isospin $I = \frac{3}{2}$.

What about the spin of a resonance? Spin is essentially angular momentum; in the case of the N* here, its spin is the sum of the orbital angular momentum determined by the orbit of the incoming pion and the intrinsic angular momenta of the pion and the nucleon (0 for pion, $\frac{1}{2}$ for nucleon). If the incident pion of momentum p is aimed at a distance r from the center of the target, that pion has an angular momentum $l = r \times p$ which, as a vector, points perpendicular to the plane of the orbit (z-axis) as in Fig. 7.8. According to the quantum theory, l becomes quantized to integer multiples of Planck's constant $\hbar$. Depending on whether the spin of the nucleon is parallel (up) or anti-parallel (down) to the direction of l, the total angular momentum j is $(l + \frac{1}{2})\,\hbar$ or $(l - \frac{1}{2})\,\hbar$. If, for example, $j = 3\hbar/2$, the neutron spin is either parallel with $l = 1$ or anti-parallel with $l = 2$.

In general, the interaction among particles depends on the angular momentum; there are instances when there is an especially strong attraction at a particular j and a particular energy. This is when a resonance occurs. One can say that, in the case of N^* ($j = \frac{3}{2}$), there is a strong attraction between the pion and the nucleon when $l = 1$ and spin is parallel which makes the pion go in circles around the nucleon before it flies away. The more circuits the pion makes, the longer the lifetime and the narrower the width of the resonance energy.

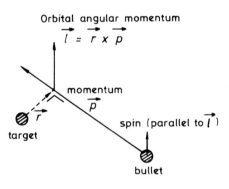

Fig. 7.8   The orbital angular momentum is perpendicular to the plane of the orbit.

But how can we tell that the spin of the N^* is $j = \frac{3}{2}$? There are two simple indications of the spin. The first is the angular distribution of the scattering which uniquely depends on the value of j. Roughly stated, the larger the value of j, the more complex the pattern of the angular distribution. The second is the size of the cross section which depends on j and the de Broglie wavelength of the incoming particle.

In general, many resonances exist in reactions of two hadrons. Beginning with the N^*, many such cases have been discovered. One can say that there are practically an infinite number of them. This leads us to believe that hadrons are compound particles and their internal excitations appear as resonances. We already had the same experience with the atoms and nuclei; the hadrons are not much different.

8

PARITY AND CONSERVATION

What is Parity?

So far I have spoken about the various quantum numbers and explained how they give rise to the selection rules for particle interactions. The conservation laws are not only useful but are seen as something that reflect the general principles of natural phenomena. Conservation laws can be absolute like that of energy or electric charge, or only approximate like that of isospin or strangeness; the former types are like the constitutions of nature.

For example, you have probably heard that "perpetual motion machines" cannot be made. It is natural for human beings to wish for a machine that creates energy from nothing, just as it is natural to wish for a flying machine; physicists, however, reject the notion of perpetual motion machine from the start. Why do they, with such conviction, treat conservation laws like some holy object to be worshipped? The explanation of this conviction is the purpose of this chapter.

Briefly, conservation laws are intimately connected with the various symmetries of the natural world. A theorem was proved by the German woman mathematician Emmy Noether that if a symmetry exists, there also exists a corresponding conservation law. Since we cannot get into mathematics we will remain with an intuitive explanation of this remarkable phenomenon. Let's start with the conservation of energy and momentum.

The above two conservation laws are thought to arise from the homogeneity of time and space, respectively. Homogeneity means that any one point is like another; the laws of physics that govern a phenomenon at a particular point in space and at a certain time are no different at any other point and any other time. This is the reason, for example, why there exists a wave that extends forever in both space and time at a fixed wavelength and frequency. And since in quantum mechanics wavelength and momentum, frequency and energy, are related by the Einstein-de Broglie relationship, it can be vaguely seen that there must be some connection between conservation laws and homogeneity.

In the same way, conservation of angular momentum is related to the isotropy of space. Isotropy means that the direction one faces at one point makes no difference to the circumstances. Homogeneity and isotropy are seen as fundamental symmetries of time and space, so it is no wonder that the conservation laws that arise from these are regarded as absolute.

As may be noticed from the above discussion, symmetry indicates that there are several equivalent points of view, and a change from one of these points to another makes no difference in the laws of nature. Thinking about conservation laws in this way makes it easy to consider a nearly conserved law such as that of isospin. Since the proton and neutron, or π^0, π^+, and π^-, have similar characteristics and can be more or less interchanged, isospin is almost conserved.

Which symmetry, then, does the conservation of electric charge correspond to? It turns out that electric charge can be thought of as a kind of angular momentum. But this angular momentum does not concern rotation in real space; it is rather a symbolic rotation concerning the quantum mechanical wave (wave function). The fact that the charge is quantized to 0, ± 1, ± 2, . . is like the case of the normal angular momentum. This sort of geometrical interpretation of electric charge is the starting point for the gauge field theories that we will speak of later.

Let us now get back to the symmetry of space-time. The symmetry of space-time is not limited to the translation and rotation discussed above; one can also think of reversal. There are two kinds of reversal; reversal in space and reversal in time. The unique characteristic of the reversals is that they are not a gradual continuous change but a sudden discontinuous change; in fact, two repeated reversals bring everything back to the original condition.

The space reversal is also called the mirror image. It is a question of looking at the right-left, or up-down, reversed image in the mirror. Although human beings are built roughly right-left symmetric, everyone knows that a right-handed person has a left-handed mirror image. Since facing different directions will not change a right-handed person into a left-handed person, the world of mirror images must be a different place from the real world.

Let us compare the above situation with the continuous symmetry mentioned before. Let's say that you have two color slides of your friend Mr. X. One picture was taken in Tokyo and the other in Osaka, but the projection of the slides reveal that both are Mr. X. Thus Mr. X is invariant under translation of coordinates by 500 kilometers. But what if the slide was put in backwards by mistake? If Mr. X had a mole on his right cheek, we would know immediately that there has been a mistake; such a Mr. X doesn't exist.

On the other hand, it is natural to assume that in fundamental units like elementary particles that do not possess a "shape", there is no distinction between right and left. In other words, the world that one sees in the mirror is "theoretically" also possible, and it obeys the same laws as our world. In this context, there is no reason why there cannot exist the alternate Mr. X with a mole on his left cheek.

The conservation law that accompanies this kind of spatial inversion is called the conservation of parity; its effect is only felt in quantum theory. This is a question of whether the hills and valleys of the quantum mechanical wave exchange positions under the inversion. For example, if a wave that moves around a point has an even number of hills and valleys (angular momentum l = even number), then a hill can be mapped into a hill under inversion; if there is an odd number of them, then the hills change into valleys. The former case is said to have parity plus and the latter minus. Parity is conserved when the parity of the initial state is the same as that of the final state in a reaction.

Just as angular momentum can be orbital or intrinsic, parity can also be orbital or intrinsic. And unlike the usual quantum numbers, parity has only two values, plus or minus; this means that the parity of a compound state is not the sum of the parities of the constituents but the product. For example, the intrinsic parity of the π-meson is minus but if two pions are together and the orbital angular momentum is zero, the total parity is plus.

The Non-Conservation of Parity

The conservation of parity, i.e., the equivalence of right and left, had been thought to be an obvious part of the symmetry of space, when, in 1957, the physics community was shocked to find that such is not the case. It began with the decay of the K-meson. The K-meson, as shown in Fig. 7.1, is one of the strange particles and comes as a pair K^+, K^0 (strangeness $+1$), and a pair of its antiparticles $\bar{K}^-$, $\bar{K}^0$ (strangeness -1); their isospin is $\frac{1}{2}$ but they decay into various particles via weak interaction. For example, in the case of K^+:

$$K^+ \rightarrow \pi^+ + \pi^0 \qquad (\theta\text{-mode})$$
$$\rightarrow \pi^+ + \pi^+ + \pi^- \quad (\tau\text{-mode})$$
$$\rightarrow \mu^+ + \nu_\mu \qquad (K_{\mu2}\text{-mode})$$

and other decays occur at fixed probabilities.

When the K-meson was observed as a V particle, it was not known that the various decay modes shown above were from the same particle; thus the names in the parentheses were given depending on the decay phenomena (p. 70, Fig. 7.5). (The name τ should not be confused with the τ lepton which came much later.) It is of course necessary that the mass be the same if the parent particles were the same, but all the other quantum numbers, such as spin and parity, must match as well.

It was eventually found that the tau (τ) particle and the theta (θ) particle had the same mass. The spin of the particles is also the same and is zero; but the parity of the tau is minus, that of theta plus. This conclusion comes from the parity of the pion which is minus.

Is it then that the tau and theta are different particles that happen to have the same mass and spin but different parity? This was the so-called "τ-θ puzzle" which gave headaches to the theorists; the mystery was solved eventually by two young physicists at Princeton's Institute for Advanced Study, C.N. Yang and T.D. Lee. They pointed out that there was not yet any evidence that parity is conserved in the weak interactions such as the beta decay; they explained that the theta and tau are the same particle but they can decay into two or three pions since parity is not conserved in its decay processes.

C.N. YANG　　　　　　　　　**T.D. LEE**

Chen-Ning Yang was born in Hefei, Anhui, China in 1922. He entered the University of Chicago after the war. At that time, physicists like Enrico Fermi and "the father of the hydrogen bomb" Edward Teller, were active at Chicago; Yang became Teller's student. The Fermi-Yang theory was produced around that time; the Ising model of ferromagnetism is also a famous work of his from this period. He then became a member of the Institute for Advanced Study at Princeton, and came under the wings of its director J.R. Oppenheimer. In 1954, he produced, with R.L. Mills, the Yang-Mills theory, pioneering non-Abelian field theories; the true physical meaning of this theory was not realized until some ten years later. He announced, in collaboration with T.D. Lee, the general theory of non-conservation of parity, which was inspired by the so-called "tau-theta puzzle". Several months later, their theory was confirmed by experiments by Madame C.S. Wu and L.M. Lederman. Yang and Lee received the Nobel Prize the following year. Their theory had a strong impact on all physicists as it made clear that nature does not respect symmetry. Yang is presently a professor at the State University of New York.

Tsung-Dao Lee was born in 1926 in Shanghai. He is slightly junior to Yang; he also came through the University of Chicago to Princeton. At Chicago, he studied under the famous astrophysicist S. Chandrasekhar. At one time, "Chandra" commuted every week to Chicago from the Yerkes Observatory in Wisconsin to lecture for Lee and Yang. Chandra once said, "Everyone in my class won the Nobel Prize." As with Yang, Lee has made many contributions in statistical mechanics and field theory. He is presently a professor at Columbia University.

Experiments were carried out right away to test their hypothesis. L.M. Lederman the present director of Fermi National Accelerator Laboratory and R.L. Garwin at Columbia University analyzed the $\pi - \mu - e$ (pion-muon-electron) decay sequence; Madame C.S. Wu's group at the same university analyzed the beta decay of (an isotope of) Cobalt. Both experiments confirmed that the parity was not conserved in the decay.

An important point that became clear from these experiments concerned the spin of the neutrino; the neutrino, produced in beta decay with the positron, is always left-handed (left helicity) and the antineutrino, produced with the electron, is always right-handed (right helicity). Here right-handed and left-handed mean that the spin is right-handed $(+\frac{1}{2})$ or left handed $(-\frac{1}{2})$ with respect to the direction of motion, just as in the circular polarization of light.

The fact that left helicity changes into right helicity when projected in the mirror can be seen from the above picture. Thus

Left helicity becomes right helicity upon reflection by a mirror.

nature does not always conserve parity. In the actual experiment, the spin of the neutrino is not directly observed; but the helicity of the neutrino causes an asymmetry in the direction in which the electron is released in the process of beta decay. In Lederman's experiment, the π^--meson, produced by a cyclotron, decays into μ^- (muon) and $\bar{\nu}_\mu$ (mu-antineutrino) and the muon then decays by emitting an e^- (electron), $\bar{\nu}_e$ (electron-antineutrino) and ν_μ (mu-neutrino); the fraction of the e^-'s that come out in the direction of the momentum of μ^- is larger than those that come out in the opposite direction.

CP Violation

We have found that the neutrino does not have a left-right symmetry, but if we back off one step we can still say that that there is a symmetry. Since the neutrino ν is left-handed and antineutrino $\bar{\nu}$ is right-handed, one can take the mirror image and also exchange particle and antiparticle at the same time. The particle-antiparticle exchange must take place for all the particles at once for the operation to have a meaning, so the electron and proton charges become reversed; in this sense, this exchange is called charge conjugation. The above simultaneous change of charge and parity is called, from their initials, CP.

It was thought, for a few years after the parity experiments, that natural laws are invariant under the operations of C and P, except in the case of the weak interaction which is, however, invariant under the combined operation CP. In fact, even if all the particles in the world were exchanged with their antiparticles, there would be no differences electromagnetically since the sign of a change is only a relative notion, and there would be no changes in the weak interactions if parity were exchanged at the same time, since then the right-handed antineutrino $\bar{\nu}$ becomes left-handed just like in the original world of the neutrino ν; in the case of the above mentioned $K^\pm$, the K^+ in the mirror would behave in the same way as the K^-.

Then physicists were dumbfounded again, in 1964, when it was discovered that there is a phenomenon in which CP symmetry is also broken. The discoverers of this phenomenon were V.L.

Fitch and J.W. Cronin of Princeton University, and it again concerns the K-meson; however, the phenomenon is even more subtle in nature than the parity violation and is harder to explain. (Fitch and Cronin won the Nobel Prize for this work.)

There is a test which is conveniently very sensitive to the violation of CP. The test involves the neutral K-mesons, K^0 and $\bar{K}^0$. These are each other's antiparticles; as in the case of the $K^\pm$, they are observed as V particles in their decay into two pions (π^+, π^- or π^0, π^0) and three pions (π^+, π^-, π^0 or π^0, π^0, π^0). These final states of decay have the same symmetry in charge, i.e., both have C equals plus, but the parity of the former is plus and the latter minus. If the decay process conserves parity, the former is the decay of a particle of CP equals plus and the latter that of CP equals minus and these must have different parent particles.

The CP equals plus particle is called K_1, and the minus one K_2. These are thought to be the following combinations of the K^0 and $\bar{K}^0$:

$$K_1 = K^0 - \bar{K}^0 \quad (C = -1, P = -1, CP = +1)$$
$$K_2 = K^0 + \bar{K}^0 \quad (C = +1, P = -1, CP = -1).$$

It can also be said that the K^0 and $\bar{K}^0$ are combinations of K_1 and K_2.

This situation is exactly like that of polarized light. A linearly polarized light can be thought of as a combination of two circularly polarized lights, and a circularly polarized light can be made as a combination of two linearly polarized lights also; neither the linearly polarized nor circularly polarized light can said to be fundamental. But when light passes through a polarizing medium, either one component of a linearly polarized light or one component of circularly polarized light could be absorbed depending on the kind of medium light is traversing; thus it is advantageous to think of light in terms of different kinds of polarization depending on the medium.

In the case of K-meson production by strong interactions it is advantageous to think in terms of K^0 and $\bar{K}^0$, which are distinguished by strangeness; conversely, it is advantageous to think in terms of K_1 and K_2, distinguished by CP, in the case of the decay by the weak interaction. Since K_1 and K_2 are different particles in decay, they may have different lifetimes; in fact the litetime of the K_1 is only $\frac{1}{100}$ of that of K_2. When initially a K^0 (or $\bar{K}^0$) beam is produced, this is a mixture of K_1 and K_2; the K_1 part decays away quickly and only the K_2 remains after a few meters.

Fitch and Cronin discovered that this remaining K_2 can also decay to two pions. Since K_2 can decay to both 2π and 3π, CP is not conserved.

Are Natural Laws Invariant under Time Reversal?

There is something called the CPT theorem; T means the time reversal operation, something like running movies backwards. It has been known for a long time that the fundamental laws of physics are invariant under time reversal. Nevertheless, the direction of time is fixed in the real world; heat flows from the higher to the lower temperature; men grow old. This problem has puzzled physicists for the past hundred years; we cannot begin to discuss this subject here.

The present problem, however, is whether the natural laws are really invariant under time reversal (T). According to the CPT theorem, the natural laws are, in general, invariant if the three operations C, P and T are carried out consecutively. If something is not invariant after CP, then T must cancel the noninvariance; this implies that T operation also breaks the invariance.

Thus none of the operation C, P, or T is an exact symmetry of nature. But since only the weak interaction breaks the symmetries, there are practically no effects in most phenomena. One can say that there is a slight blemish in nature but it is hidden quite well.

Let me make it clear, once again, that the asymmetry seen in the everyday world and the asymmetry of nature are different things. Most people are right-handed; the DNA's in living organisms are in general right-handed; the solar system is made of nucleons and electrons and not of antinucleons and anti-electrons. But this does not mean that a world which is a P or C conjugate of our own world cannot be created in principle. Once such a world is created, it would behave in the same way as our own world, and it would be difficult to tell the difference. For example, in a reversed world, most people would be left-handed, cars would be driven on the right side (or the left side, depending on the country); the earth would reverse east and west, and the sun would rise from the east (the original west!). But could the inhabitants of this world know the difference?

Even if the natural laws are right-left symmetric, one must start out from certain conditions to realize the world. They are called initial conditions, which are in principle arbitrary; one may say that the initial conditions are governed by chance. The real world which is made up of countless particles is incredibly complex; thus, if the first molecules created happen to be left-handed, all the following chemical reactions may be influenced by them, and only left-handed molecules may be made. The initial conditions determine the actual asymmetry.

Sheeplike dependance on outside forces is not merely a human characteristic but is also applicable to physical phenomena. Thus, there are many cases in which symmetry is broken naturally (or spontaneously), but the details must wait until Chapter 19.

9

COMPOSITE MODELS
OF HADRONS

What is Expected from Fundamental Particles

The Gell-Mann-Nakano-Nishijima Law made clear that the reactions of hadrons are governed by definite rules concerning the quantum numbers isospin and strangeness carried by each hadron. On the other hand, it was found that hadrons have many resonances (excited states), and thus behave somewhat like compound particles such as the nuclei and atoms. From the above two facts in combination, it would seem natural to assume that hadrons are made of some other fundamental particles, and these fundamental particles carry the quantum numbers.

It is not easy, however, to bring in new particles, even if it seems to be the obvious thing to do from hindsight. Once one gets accustomed to the idea that the neutrons and protons make up the nuclei, and that the pion mediates the nuclear force, then it takes some courage to break this image when there is no evidence that strongly contradicts it. Researchers usually avoid new theories unless absolutely necessary. This is in accordance with the tradition of natural science which values evidence above all.

On the other hand, history has taught us, many times over through bitter lessons, that superficial descriptions of phenomena are not the only goals of natural sciences. Fortunately, a tradition exists in Japan which emphasizes constructive modes of thinking, as exemplified in the Yukawa theory. S. Sakata and M. Taketani, collaborators of Yukawa, consciously emphasized this point and expounded it as a methodology of particle physics research. They influenced, to a large degree, physicists who came after them, and their efforts led indirectly to many successes. We will speak of these matters later; but let us now start with the Sakata model.

The Composite Model of Fermi and Yang

Soon after the existence of the pion was established, E. Fermi and C.N. Yang presented an interesting theory. Yang was still a student at the University of Chicago at that time; it was some time later that he and T.D. Lee predicted the non-conservation of parity, thereby earning the Nobel Prize. The Fermi-Yang theory pointed out that quantum numbers work out correctly if one thinks of the pion as a composite particle made up of a nucleon and an antinucleon. Thus, one can make combinations out of proton (p), neutron (n), antiproton ($\bar{p}$), and antineutron ($\bar{n}$):

$$\pi^+ = p\bar{n}, \qquad \pi^- = n\bar{p}, \qquad \pi^0 = (p\bar{p} - n\bar{n}).$$

If the above are assumed as the chemical equations for the pions, the isospin as well as the electric charge comes out right; this is because p, n, $\bar{p}$, and $\bar{n}$ have $I_z = \frac{1}{2}, -\frac{1}{2}, -\frac{1}{2}, \frac{1}{2}$ respectively, whereas π^+, π^0, and π^- have $I_z = 1, 0, -1$. The equation for π^0 above indicates, just as in the case of K_1 and K_2 (p. 87), a superposition of waves. In the present case, the subtraction (minus sign) is necessary to make a state of isospin length $I = 1$; but we will not go into the mathematical reasons why this is so.

This takes care of the isospin of the pion but what about the spin? We know that the spin of the pion is 0 and its parity is minus. The nucleon has spin $\frac{1}{2}$, so if we combine two nucleons of opposite spins, their total spin can indeed be made equal to 0. And what about the parity? This works out too. Antiparticles have opposite parities from particles, so the product of their parities is minus.

	p	n	$\bar{p}$	$\bar{n}$	π^+	π^-	π^0
I_z	$\frac{1}{2}$	$-\frac{1}{2}$	$-\frac{1}{2}$	$\frac{1}{2}$	1	-1	0
Spin	$\frac{1}{2}$	$\frac{1}{2}$	$\frac{1}{2}$	$\frac{1}{2}$	0	0	0
Parity	$+$	$+$	$-$	$-$	$-$	$-$	$-$

Fig. 9.1 Fermi-Yang quantum numbers make sense when the π-meson is thought to be a compound particle of a nucleon and an antinucleon.

The "Elementary" and "Compound" in the World of Elementary Particle Physics

Is the pion really a composite state of the nucleon and the anti-nucleon? The primary proof of the compound nature of the atom and the nucleus is the fact that these can be taken apart into their constituents. In the same way, can we take the pion apart? That is, when we put in an energy X from the outside, can we get the reaction:

$$X + \pi^- \to \bar{p} + n?$$

It is not as easy to prepare a sample of pions as to prepare a sample of nucleons; but the inverse reaction of the above can also take place. In the latter case, the energy X will be released in the form of various particles (γ, π^0, . . .). But then such a reaction is no different from the reactions we have been talking about all along. By this argument, we may say that the photon is a composite state of an electron and a positron, since the former can decay to a pair of the latter. We see that a confusion ensues if we try to define exactly what is "elementary" and what is "compound" in the world of "elementary" particles where interchanges of particles occur freely. In the cases such as the hydrogen atom, the binding energy is small compared to the masses of the constituent particles (proton and electron), which means that the binding is weak, and the constituent particles retain their original characteristics. In the case of the pion, on the other hand, the binding energy is large and not much different from the mass of the constituent nucleons; so the nucleons inside the pion may look like something completely different when seen under a (hypothetical) microscope.

In any case, we must admit that the distinction between elementary and compound particles is, to a certain extent, a matter of convention. If we could, however, minimize the number of fundamental particles and derive the properties of all other particles from these few particles, then we would have made enormous progress in the minimization of postulates. It is the purpose of elementary particle physics to explain the complex from the simple; but I want to point out that when one begins to explore the subject as a practical problem, the differences between elementary particles and compound particles become less and less like the difference between black and white.

I do not know how seriously Fermi and Yang took the composite model of the pion. Perhaps Fermi wanted to try something unconventional since he, at that time, apparently did not trust the quantum field theory although it was the basis of the Yukawa theory, among other things. Presumably, the reason he did not trust quantum field theory was because the theory often gave nonsensical answers except when it was applied to the simplest phenomena. This difficulty, generally called the problem of the infinite self-energy, had at that time just been resolved by Tomo-

---

ENRICO FERMI (1901–1954)

Fermi was an all-around Italian physicist who made many key contributions to both experimental and theoretical physics. He was born in Rome and took his degree at Pisa; he became a professor at the University of Rome at the age of 26. He won the Nobel Prize in 1938 for his experimental work using neutrons. In theoretical physics he is famous especially for Fermi-Dirac statistics (1927) and the beta decay theory (1934). Soon after he came to Columbia University, in the United States, to flee from the Fascist government in Italy, he heard the news of the discovery of nuclear fission in Germany and immediately began experiments; this was the beginning of the nuclear power research. On December 2, 1942, the nuclear reactor built at the University of Chicago under Fermi's guidance reached critical mass and proved the feasibility of chain reactions. After the war, he returned to pure physics; he built a cyclotron at the University of Chicago and began experiments on pions. He unfortunately passed away at the age of 53.

Fermi was a born leader and many of the traditions he created, as well as many stories concerning him, survive at the University of Chicago. He lectured lucidly and brilliantly in class, but he was also famous for giving very difficult qualifying examinations for doctoral candidacy. I once wrote in a letter to a friend after I first saw Fermi give a lecture: "Fermi was like a kabuki actor on stage."

naga, Schwinger, and Feynman; but it was not yet clear that they had found the complete solution.

It is important to distinguish between the mathematical method of the description of elementary particles and the problem of what kinds of particles do actually exist. There is no reason not to pursue the latter even if the former is incomplete. It was in this spirit that S. Sakata pushed the Fermi-Yang model one step further and attempted an interpretation of the strange particles.

The Sakata Model

According to the theory called the Sakata model, the lambda particle (Λ), in addition to the proton (p) and the neutron (n) of Fermi and Yang, make up the three fundamental baryons. All other hadrons, be they mesons or baryons, are compound particles made up of this Sakata trio. For example:

$$\begin{array}{lll}
\pi^+ = p\bar{n}, & \pi^- = n\bar{p}, & \text{(Fermi-Yang)} \\
K^+ = p\bar{\Lambda}, & K^0 = n\bar{\Lambda}, & \\
K^- = \Lambda\bar{p}, & \bar{K}^0 = \Lambda\bar{n}. &
\end{array}$$

So, in the case of mesons, the theory is an extention of the Fermi-Yang model as shown above; but the case of baryons is more complicated. Since baryons obey Fermi statistics and have half-integer spin, they must be composed of a combination of 3 (or 5, 7, . .) particles from p, n and Λ. For example, the sigma particle (Σ) is a combination of Λ and π, the xi particle (Ξ) is a combination of Λ and K, so

$$\begin{array}{lll}
\Sigma^+ = & \Lambda\pi^+ & = \Lambda p\bar{n}, \\
\Xi^0 = & \Lambda\bar{K}^0 & = \Lambda\Lambda\bar{n}.
\end{array}$$

The great advantage of the Sakata model is the fact that the origin of the GNN (Gell-Mann-Nakano-Nishijima) law is beautifully explained. Thus, the proton (p) and the neutron (n) each carries isospin $\pm\frac{1}{2}$ and the lambda particle carries strangeness. These quantum numbers are nothing other than the labels that distinguish these three fundamental particles. When we discussed the relationship between symmetry and conservation, we attributed the (near) conservation of isospin to the similarity of protons and neutrons; it is natural to apply the same arguments to the trio, proton, neutron and lambda.

M. Ikeda, S. Ogawa, Y. Ohnuki, and independently, Y. Yamaguchi were the people who actually carried out the above idea. In mathematical language, the symmetry among three objects is called SU_3 symmetry. SU is an abbreviation for special unitary group and indicates that the transformations that mix three states make up a group called SU_3. In the same way, the transformations that mix two states (for example, p and n) form a group called SU_2, and this symmetry is nothing other than the isospin symmetry.

To distinguish between two different states, each state must be given a different quantum number; for example the component of isospin I_z could be $\pm \frac{1}{2}$ (up or down). To distinguish the third state from the first two, another quantum number is

---

SHOICHI SAKATA (1911-1970)

Sakata was born in Tokyo. He became a student and later a collaborator of Yukawa at Kyoto University and contributed to the early development of the meson theory along with Mitsuo Taketani, Minoru Kobayashi and others; he then became a member of the Nishina-Tomonaga group at Riken (Institute for Physical and Chemical Research), after which he went to the Nagoya University to establish his own style of research. The Taketani-Sakata two meson theory was introduced in 1942, and the Sakata model in 1956. He also influenced young researchers through his writings and lectures on the critique of methodology based on dialectic materialism. For example, he claimed that there are infinite layers of elementary particles; this helped reduce the psychological barrier, felt by many researchers against introducing new particles. It was under these circumstances that Maki's four quark theory and the Kobayashi-Maskawa six quark theory (Chapter 18) were born. It may not be an exaggeration to say that particle theory has been proceeding according to Sakata's scenario.

needed. Strangeness S fits the bill exactly. Since the proton and neutron each has $I_z = \pm\frac{1}{2}$, $S = 0$, and the lambda has $I_z = 0$, $S = -1$, specifying I_z and S is enough to indicate which of the three we mean (Fig. 9.2).

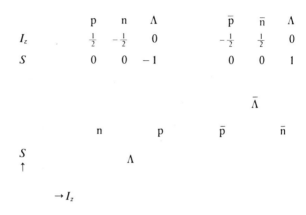

Fig. 9.2   Quantum numbers of two Sakata triplets.

In general, a group of n particles can be distinguished by giving them $n - 1$ different quantum numbers.

But the choice of the quantum numbers is largely arbitrary. For example, in the case of the triplet (p, n, Λ), if the three states really have the same physical characteristics, one could, for example, distinguish the neutron from the lambda with I_z, and distinguish the proton from the neutron and the lambda with S. But in reality, the significance of the quantum numbers of the GNN equation comes from the fact that of the trio (proton, neutron and lambda), the proton and the neutron are especially alike; so the GNN quantum numbers are not just historical in origin.

In the previous chapter we said that the isospin symmetry (SU_2) not only conserves I_z but also gives relationships between different reactions; the same can be said about SU_3. Thus if one ignores the slight differences among the proton, neutron, and lambda, the reactions that involve different combinations of them should satisfy definite relationships with each other. This is one of the powerful results of the SU_3 symmetry theory.

Furthermore, the arguments of SU_3 symmetry can tell us what kind of compound states can be formed from the three fundamental particles (and antiparticles) when a given number of these particles are combined. For example, if compound states of two particles are made from the fundamental triplet (p, n, Λ), there will be $3 \times 3 = 9$ kinds (pp, pn, np, . . . etc.). But if the symmetry under the exchange of the two particles is taken into account, these nine divide into a symmetric group of six and an antisymmetric group of three, and these two groups may have different physical characteristics; for example, they may have different masses (Fig. 9.3).

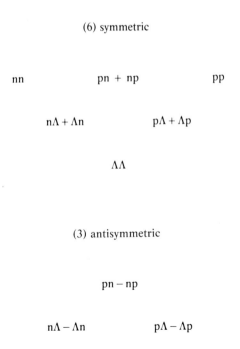

(6) symmetric

nn   pn + np   pp

nΛ + Λn   pΛ + Λp

$\Lambda\Lambda$

(3) antisymmetric

pn − np

nΛ − Λn   pΛ − Λp

Fig. 9.3   Compound states of 3×3.

In the same way, the number of states that come from the combination of a particle and an antiparticle is also 9, but in this case, by symmetry arguments, the states divide into a group of eight and a group of one (Fig. 9.4). Expressing the triplets (p, n, Λ) and ($\bar{p}$, $\bar{n}$, $\bar{\Lambda}$) as 3 and $\bar{3}$ respectively, the above situation may be expressed in the form

$$3 \times 3 = 6 + \bar{3},$$
$$3 \times \bar{3} = 8 + 1.$$

Here the $\bar{3}$ on the right-hand side means that the corresponding compound states behave like ($\bar{p}$, $\bar{n}$, $\bar{\Lambda}$) as far as the quantum numbers are concerned.

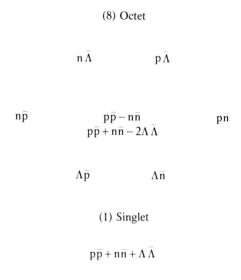

(8) Octet

$n\bar{\Lambda}$ $p\bar{\Lambda}$

$n\bar{p}$ $p\bar{p} - n\bar{n}$ pn
$p\bar{p} + n\bar{n} - 2\Lambda\bar{\Lambda}$

$\Lambda\bar{p}$ $\Lambda\bar{n}$

(1) Singlet

$p\bar{p} + n\bar{n} + \Lambda\bar{\Lambda}$

Fig. 9.4 Compound states of $3 \times \bar{3}$.

To appreciate the power of the SU_3 symmetry, the meson family should be examined. If the mesons are considered as belonging to the $3 \times \bar{3}$, the octet (8) in Fig. 9.4 corresponds to the octet in Fig. 7.1 on p. 67 (π, K, η). The singlet (1) seems to correspond to the meson named η'; the fact that this particle has a much larger mass than the other eight agrees with the above method of classification. Also, the mass differences among the octet (for example $K^+ = p\bar{\Lambda}$ is heavier than $\pi^+ = p\bar{n}$) can be explained simply by the fact that the lambda particle (Λ) is heavier than the nucleon.

Unfortunately, the Sakata model stumbles when the baryons are considered. As already stated, the baryons, like the mesons, also come in octets (Fig. 7.2). The Sakata model has singled out the p, n, and Λ as special baryons which are fundamental, but the five other particles of the group (Σ, Ξ) are not especially different from the p, n, and Λ; not only is this unnatural; a group of five does not exist in the SU_3 symmetry.

One way out of the situation would be to take the known resonances with the same isospin and strangeness as the proton, neutron, and Λ and consider these as a part of the octet along with the sigma (Σ) and the xi (Ξ).

But such an attempt did not succeed, and the quark model of Gell-Mann and Zweig replaced the Sakata model. We already spoke of quarks in Chapters 2 and 3. In the next chapter, I will explain how well the quark model can describe the characteristics of hadrons.

10

THE QUARK MODEL

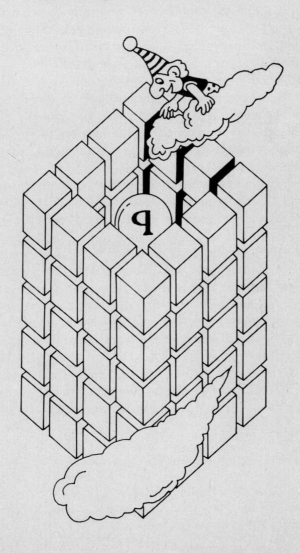

How does the Quark Model Differ from the Sakata Model?

The difference between the Sakata model and the quark model is that in the case of the quark model the fundamental particles are postulated as one layer beyond the known hadrons. Thus, the hadrons are all compound particles and, as such, are all on an equal footing. The problems of the Sakata model, arising from the fact that the proton, neutron and lambda are treated as special among the baryons, now dissappear. However, one must then take the seemingly unnatural position that the baryons are compound particles made up of three quarks.

Actually, the quarks themselves are not much different in their characteristics from the proton, neutron, and lambda of the Sakata theory. This is so because the quarks, like the Sakata triplet, were conceived as the carriers of isospin and strangeness — except for one sensational difference: they carry a fractional charge.

The quark model was introduced independently by M. Gell-Mann and G. Zweig in 1964. Gell-Mann called the triplet of new particles *quarks*, and Zweig called them *aces*, like the aces in playing cards.

Gell-Mann claims to have taken the name quark from James Joyce's novel *Finnegans Wake*. These names, quark or ace, merely indicate their authors' witticisms and there is no need to probe into their meanings. The reason the name quark has become popularly accepted must be because of the combination of the mystical sound of the word and the stature of its author.

The three quarks of Gell-Mann and Zweig are conventionally denoted by u, d, and s. These correspond to the p (proton), n (neutron), and Λ (lambda) of the Sakata model. Thus:

particle $\quad$ (u, $\quad$ d, $\quad$ s) $\longleftrightarrow$ (p, n, Λ)

charge $\quad$ $(\frac{2}{3}, -\frac{1}{3}, -\frac{1}{3})$ $\longleftrightarrow$ (1, 0, 0)

u and d indicate the directions of isospin, up and down; s indicates strangeness. Note that the quark charges are displaced by $\frac{1}{3}$ as compared to the charges of the Sakata triplet. As a result, the sum (or average) of the charges of three quarks is zero, but the differences between the charges of various quarks are the same as in the case of pnΛ; it should be said, in fact, that the quark model was conceived with the above requirements in mind.

	p	n	Λ	u	d	s
I_z	$\frac{1}{2}$	$-\frac{1}{2}$	0	$\frac{1}{2}$	$-\frac{1}{2}$	0
strangeness	0	0	-1	0	0	-1
electric charge	1	0	0	$\frac{2}{3}$	$-\frac{1}{3}$	$-\frac{1}{3}$
spin	$\frac{1}{2}$	$\frac{1}{2}$	$\frac{1}{2}$	$\frac{1}{2}$	$\frac{1}{2}$	$\frac{1}{2}$

Fig. 10.1 Comparison of fundamental particles of the Sakata and the quark models.

MURRAY GELL-MANN

He was born in 1929 in New York and received his degree from the Massachusetts Institute of Technology (MIT). He is one of the key figures in the formation of postwar particle theory. He became an instructor at the University of Chicago, and was promoted every year until he left to become professor at the California Institute of Technology. The Gell-Mann-Nakano-Nishijima law (1953) dates from this period. Among his other accomplishments are the Gell-Mann-Low renormalization group equation, the Feynman-Gell-Mann V minus A theory of weak interaction, the Eightfold Way (introduced independently by Y. Ne'eman), the

quark model (introduced independently by G. Zweig), the theory of current algebra, and so forth. Among these, the concept of renormalization group came to be recognized for its real significance some 20 years later when it began to be applied to gauge fields and statistical mechanics. He won the Nobel Prize in 1965. He is also well versed in linguistics; he is the author of such technical terms as strangeness, quark, color and so forth. He could be said to be the one who Americanized, both in word and deed, particle theory.

In the Sakata model, the meson was a combination of particle and antiparticle, and this is also the case with the quark model; we merely replace p, n, Λ with u, d, s. The shift of $\frac{1}{3}$ in the quark charges is cancelled out by the shift of $-\frac{1}{3}$ in the antiquark charges.

For simplicity, from now on, the quark and antiquark will be indicated by q and $\bar{q}$ respectively; the u, d, and s quarks will be distinguished as q_1, q_2, and q_3. Then the mesons M will be expressed as

$$M = q_i \bar{q}_j \qquad (i, j = 1, 2, 3)$$

in chemical equation form. To see the correspondence between these symbols and the various mesons, compare Fig. 7.1 on p. 67 and Fig. 9.4 on p. 100. As before, the $3 \times 3 = 9$ combinations will break up into an octet and a singlet.

Baryons in the Quark Model

Then what about the baryons? The difference between this model and the Sakata model becomes clear here. The fact that the GNN law follows from the SU_3 symmetry was pointed out by M. Gell-Mann and by Y. Ne'eman independently after the introduction of the IQQ theory; their work was not based on composite models, like the Sakata model but was based only on the mathematical symmetry. They reasoned that the known mesons and baryons are members of the octet of SU_3. That is why Gell-Mann called his theory the Eightfold Way. (The Eightfold Way is the path to Nirvana in Buddhism.) But there is no reason for the hadrons to form only octets; rather, the triplet is the most fundamental multiplet of the SU_3 symmetry. All other multiples, or groupings, can be made from the triplet; this triplet is the quarks.

Now let me explain how the baryon octets are formed. To this end, first recall the rules for constructing compound states. There are $3 \times 3 = 9$ combinations of two out of three quarks, which divide into a sextet and a triplet (p. 99). Among these the triplet corresponds to the antisymmetric combination of two different q's, and their quantum numbers (I_z, charge Q, etc.)

are the same as those of the antiquarks $\bar{q}_1$, $\bar{q}_2$, and $\bar{q}_3$. Thus they are, as far as the SU_3 symmetry is concerned, the same as the antiquarks $\bar{q}$. Thus, an octet can be made of (qq) and q just as we made the meson octet from q and $\bar{q}$; the baryons are just such an octet:

$$B = q_i (q_j q_k) = q_i q_j q_k.$$

The details of the combinations are shown on Fig. 10.2. Of course there are $3 \times 3 \times 3 = 27$, ways to combine three q's so multiplets other than octets can also be made. For example, the decuplet (group of ten) also exists and its diagram is shown in Fig. 10.3.

d(du)   u(ud)

d(ds)   u(ds) + d(us)   u(us)
               u(us) + d(ds) − 2s(ud)

s(ds)   s(us)

Fig. 10.2 Formation of the baryon octet (8).

n^0 = d(du), P^+ = u(ud)
Σ^- = d(ds), Σ^0 = u(ds) + d(us)
Σ^+ = u(us), Λ^0 = u(us) + d(ds) − 2s(ud),
Ξ^- = s(ds), Ξ^0 = s(us)

$I_z = \frac{3}{2}, S = 0$ Δ^- Δ^0 Δ^+ Δ^{++}

$I_z = 1, S = -1$ Σ^* Σ^{*0} Σ^{*+}

$I_z = \frac{1}{2}, S = -2$ Ξ^{*-} Ξ^{*+}

$I_z = 0, S = -3$ Ω^-

Fig. 10.3 The baryon decuplet (10).

The fact that this decuplet exists as the resonances of baryons (with spin $\frac{3}{2}$) became the proof of the SU_3 symmetry. Among these, the delta particle (Δ) is a $S = 0$, $I = \frac{3}{2}$ hadron, in other words, the 3-3 resonance of πN (pion and nucleon) mentioned in a previous chapter. The next two states, $S = -1$, $I = 1$ and $S = -2$, $I = \frac{1}{2}$, already had candidate particles when the Eightfold Way was announced, but the last state with $S = -3$ and $I = 0$ was yet unknown. Looking at the known particles under the three labels, we find that the mass is increasing evenly from label to label in steps of 140 to 150 MeV. (The mass increases toward the bottom.) Applying this regularity to the last particle, the mass is predicted to be around 1680 MeV.

The Discovery of the Ω^- as Predicted

It was in 1963 that this unknown particle Ω^- (omega minus) was captured in its production and decay in a cloud chamber at Brookhaven National Laboratory by Samios *et al*.

The mass was exactly as predicted. Drawing conclusions from only one example is very dangerous and should in general be avoided; but the existence of Ω^- is not only the proof of applicability of SU_3 symmetry. If a reaction is forced between the baryon octet and the meson octet, resonances can be also in principle form a multiplet of 27. It is possible to think of, say the 3-3 resonance as part of this group, but in this case the $S = -3$ state must have $I = 1$, or come in three kinds. Thus SU_3 leaves open many possibilities, and how to choose them must be determined experimentally.

The following is a list of what the SU_3 symmetry theory tells us.

(1) Hadrons, in general, come in multiplets of one, eight and ten; within each multiplet, the members have the same spin, their masses are relatively close, and they display a pattern of strangeness and isospin that obeys the GNN law.

(2) Though the differences in masses within a multiplet indicate that the symmetry is not perfect, the breaking of symmetry also occurs in an orderly way; the mass differences

obey something called the Gell-Mann-Okubo mass formula. The decuplet case above is an example of this orderliness of mass differences; in essence, the Gell-Mann-Okubo formula expresses the relationship between mass, strangeness and isospin.

(3) A multiplet is distinguished by other similarities in such characteristics as the reaction cross sections and magnetic moments of the member particles.

The confirmation of the SU_3 symmetry by the discovery of the Ω^- led quickly to the introduction of the quark model. We spoke earlier about the fact that baryons are made of three quarks according to the quark model. The combination of three quarks, however, does not produce a multiplet of 27 ($3 \times 3 \times 3 = 27$ is broken into $1 + 8 + 8 + 10$). Thus the quark model further limits the allowed SU_3 multiplets, in accordance with what is observed.

The quark model can also explain the Gell-Mann-Okubo mass relation. One may argue that the mass differences in hadrons are mostly due to the mass differences of the constituent quarks; for example, the s quark is heavier than the u and d quarks. Then the strange particles should be heavier in proportion to the number of s quarks they contain. The mass difference of the p and n can also be understood if the d is slightly heavier than the u.

Can we get more ambitious and compare the meson and baryon masses? Mesons are $q\bar{q}$ and the baryons are qqq, and the baryons are indeed heavier than the mesons, however, the ratio of the masses is not 2 to 3. Even among the mesons, the mass differences among π, K, and η are about as large as the average mass of these particles. The quantitative aspects are not that simple.

In the early days of the quark theory, there was considerable doubt as to how seriously one should take quarks as reality. Even the originator of quarks said that these may only be mathematical symbols to express the SU_3 quantum numbers. If a quark were able to exist by itself, such doubts would disappear, but a free quark has yet to be found. Nowadays there probably are very few people who still think of quarks as just mathematical symbols, because by proceeding on the assumption that quarks are real particles, the quark theory has brought success after success.

To achieve these successes however, the quark model has had to undergo a process of evolution in various respects; let us now follow this development.

11

EVOLUTION OF THE
QUARK MODEL

Making Quark Compounds

Quarks are objects that were introduced to explain the SU_3 symmetry, but it took a long time for them to be recognized as full-fledged particles. This is understandable if one recalls the long process from hypothesizing atoms to the recognition of their existence. In fact, some people may still doubt the existence of the quark. The primary reason for this doubt is that quarks cannot be "seen". To be able to justify treating quarks in the same way as the other elementary particles, the theoretical tests other than directly "seeing" them would be necessary; and we must examine the characteristics of quarks in detail, and refine the theory if need be.

According to the Gell-Mann-Zweig theory, quarks are a triplet of spin $\frac{1}{2}$ fermions that carry SU_3 quantum numbers. (fermions, see p. 46). In other words, they are particles similar to leptons. Since hadrons are compounds of quarks, not only their isospin and strangeness but also their spin should be determined by the way the quarks are combined; and one should also be able to predict the properties of the excited states (resonances) of the hadrons. Just as the numerous resonances that exist in nuclei and atoms revealed the nature of the interactions involved by the regularity of their energy spectrum, the spectra of hadrons should also be an important indication of the nature of interaction of quarks.

Let us now construct a quark compound model based on an analogy with the theories of nuclei and atoms. Since hadrons are made up of quarks of about the same mass, it is more appropriate to compare them with nuclei than with atoms. Thus, the meson $q\bar{q}$ corresponds to the deuteron (heavy hydrogen) pn (proton-neutron), and the baryon qqq corresponds to the triton pnn or the helium–3 ppn. Deuteron and triton are isotopes of hydrogen (p); the helium–3 is an isotope of the normal helium–4 (ppnn). If we replace the proton and neutron with the u and d quarks, the triton (t) and helium–3 nuclei respectively, correspond to the composition of neutron and proton in the quark model. Thus:

$$t = pnn, \qquad He^3 = ppn,$$
$$n = udd, \qquad \quad p = uud.$$

In any case, the starting point for understanding the structure of nuclei is the assumption that a short range attractive force acts between nucleons which are bound relatively weakly; thus, the binding energy is relatively small in comparison with the rest energy of the nucleon (in the order of ten percent), and therefore the kinetic energies of the nucleons are small as well. The attractive force does not distinguish between a proton and a neutron and also does not depend on the relative direction of the spin. These properties determine the spectrum of the various energy states of the nucleus for the most part. The exact description of the nuclear spectrum is determined by adding corrections to the above due to the mass and charge differences between proton and neutron, the existence of spin-dependent forces and so forth.

Why not try the same thing with the quark model? Anybody might think of that. In fact such ideas were introduced by F. Gürsey and L.A. Radicati and by B. Sakita independently and became popular under the name SU_6; but actually this kind of reasoning is unconvincing to those who know the various principles of physics, and it takes some courage to state it.

The reason for this is that if hadrons are weakly bound states of quarks, then the quark is obviously lighter than the hadron and it should be easy to take apart a hadron into its constituent quarks. For example, if we took apart a pion into two quarks, quarks lighter than the pion should fly out; but the pion is the lightest of the particles that interact strongly. From the fact that single quarks have not been found, we must conclude that the quark itself is much heavier than the hadron, and, at the same time, the binding energy of hadrons must be correspondingly large to cancel the large rest energy. This is like having dropped a heavy stone into a deep well; it is difficult to take out the stone. But such behavior of potential energy runs into difficulties with the theory of relativity unlike Yukawa's nuclear force.

Analogy with Nuclei

Ignoring the above mentioned difficulties, let us try out the analogy with the nuclei. First the mesons; if the interquark forces do not depend on spin, the energy will depend only on the relative motion of the quarks. In general, the energy rises as the angular momentum l increases from 0 to 1, 2, 3 and so forth, so the lowest state has $l = 0$. In the case of the $l = 0$ state, the total spin is 0 or 1 depending on whether the spins of the two quarks, each of magnitude $\frac{1}{2}$, are parallel or antiparallel. The parity here is minus, and so the state can be designated as either 0^- or 1^- The above argument is exactly the same as that given in the discussion on the Fermi-Yang model (p. 92). The fact that, depending on the combinations of $q\bar{q}$, an octet and a singlet of SU_3 can be formed has already been explained. The 0^- mesons correspond to the pseudoscalar multiplet which include π, K and η (eta). The 1^- particles are called vector mesons and correspond to Ω (omega), ρ (rho), K^*, ϕ (phi) and other resonances. These are mesons that come next in energy above the 0^- mesons, and they decay into two 0^- mesons.

Let us now turn to the baryons. The lowest lying state should be when the orbital angular momentum of the three quarks is zero; then the total spin should be either $\frac{1}{2}$ or $\frac{3}{2}$ depending on the combination of the three spins. The former must be the baryon octet and the latter the decuplet. But why is there a correlation between spin and SU_3?

Here the principles of quantum mechanical statistics show their power. According to the principles of statistics (p. 45), spin $\frac{1}{2}$ particles are fermions and must obey Fermi statistics. That is to say, the compound state of quarks must be antisymmetric under the exchange of any pair of the quarks. Since spin and the SU_3 label must be changed simultaneously in the exchange operation, there must be a relationship between them in order to maintain the antisymmetry property.

The SU_6 theory of Gürsey, Radicati, and Sakita, mentioned previously, is the actual realization of the above program. They assumed that there is a symmetry among the six states of the quark distinguished by the SU_3 label (u, d, or s) and spin (up or down); hence the name SU_6.

According to the SU_6 theory, it is possible for the baryons to have an octet of spin $\frac{1}{2}$ and a decuplet of spin $\frac{3}{2}$, but one must assume a very odd thing. Quarks which were assumed to obey the Fermi statistics now seem to obey the Bose statistics. To show this, let us take the Ω^- (p.108) as an example. Since the Ω^- has spin $\frac{3}{2}$ and strangeness -3, it is thought to be a state in which the spins of three s quarks are aligned. But then the quarks are symmetric under exchange, which is contrary to the demands of Fermi statistics. But if one forgets about this contradiction and assumes Bose statistics, not only the Ω^- but the properties of all the baryons (for example, their magnetic moments) can be explained rather well.

The SU_6 theory was successful on the surface, but it had many unsatisfactory points in the underlying principles. For example, in the case of the magnetic moments just mentioned, the theory tells you to add up the magnetic moments of the three quarks assuming that each quark has one-third the mass of the nucleon; but this would mean that there is no binding energy, and it would be as easy to take a nucleon apart into quarks just as it is easy to take a nucleus apart into nucleons. But such a process has not been seen. So the quark that obeys Bose statistics like bosons (p. 46) becomes less and less like a real particle; this must be resolved somehow.

Putting aside the problem of the binding energy for the moment, let us now address ourselves only to the problem with statistics. The first suggestion to solve the problem was made by O.W. Greenberg; it was a direct approach solution in which quarks were assumed not to obey the regular statistics. In particular, the quarks obey the parafermi statistics in which up to three particles are allowed to be in the same state as opposed to the Fermi statistics in which only one is allowed. Such statistics were known theoretically but this was the first time it was used in an application to a real problem.

But this solution only treats the symptoms; if it is there only to solve the baryon problem, it is not very convincing. We need other independent tests; fortunately there is one. In a reaction that creates a $q\bar{q}$ pair, for example meson creation from e^+ and e^-, the parafermi statistics implies that the reaction probability, or the cross section, will be three times the normal value; in other words, parafermi statistics is equivalent to assuming that there are three times as many kinds of quarks than had been assumed heretofore.

Quarks Come in Colors and Flavors

Then why not simply increase the number of quarks by a factor of three? Three times means that each of the u, d and s quarks comes in three different kinds; these have the same SU_3 quantum numbers and are not distinguishable by these old quantum numbers alone. So it's like having three different SU_3 triplets. One can write them out like so:

u_1 d_1 s_1,

u_2 d_2 s_2,

u_3 d_3 s_3.

This kind of idea was conceived, independently, by M.Y. Han and Y. Nambu, by Y. Miyamoto, by A. Tavkhelidze, and several others. According to this theory, there are nine quarks in all, and they are distinguished by specifying a horizontal label and a vertical label simultaneously. The first corresponds to the SU_3 we had before, but the latter must be a new kind of SU_3. This new quantum number came to be called color later on. As opposed to color, the old quantum number is called flavor.

The color quantum numbers are called by the three primary colors red, green and blue, or red, yellow, and blue. The old labels u, d, s, however, do not particularly correspond to flavor; furthermore, there are now more than three flavors, c and b, as mentioned before and which will be discussed later. The present author has suggested using, as a translation, "species" for color and "family" for flavor as in taxonomy of animals and plants.

FLAVOR (FAMILY)			
up	u_1	u_2	u_3
down	d_1	d_2	d_3
strange	s_1	s_2	s_3
charm	c_1	c_2	c_3
bottom	b_1	b_2	b_3
	Red	Blue	Green

COLOR (SPECIES)

Fig. 11.1   Classification of quarks according to color (species) and flavor (family).

The Nine Quark Model

Let us get back to the problem of the hadron; what kind of hadrons can be made from the nine quarks? Since the number of quarks has increased, the number of baryons should increase also; but we had enough quarks already with the three flavors. Of course, in the case of the baryon, the Fermi statistics must be obeyed.

The solution to the above problem is that all hadrons are in a "colorless" or "white" state. White means that all three primary colors are equally mixed. In case of baryons, each of the red, green and blue quarks is used, as in the combination $q_R q_G q_B$, and then an appropriate average over the order of R (red), G (green), and B (blue) is taken. In the case of the mesons, the q and $\bar{q}$ of the same color are always combined, say $q_R \bar{q}_R$, and this is averaged over all colors.

Thus the difficulty with statistics is solved by assuming that all hadrons are colorless, but if the resolution of this problem was the only aim of the principle, one would have to say that it is much too artificial and *ad hoc*. On the other hand if there was some good reason to forbid the colored states that is to say, states with an unrestricted mixture of the three colors, the kinds of hadrons would increase enormously.

If there is indeed such a reason, it seems natural to assume that it is dynamical in origin. That is to say, there do exist colored states but they are much heavier than the colorless state and present experiments are unable to produce them. This is easy to understand if one compares the situation to the atomic case. Here "white" would mean the electrically neutral state where the plus charges and the minus charges have cancelled each other out, or in other words, non-ionized atoms. The colored states are ions that carry some net charge. Clearly, the ions have higher energy and tend to get back to neutral atoms. This is because an attractive Coulomb force acts between opposite charges.

Let's apply this analogy to color. Can we make a theory in which there is a Coulomb-like force, which is proportional to the quark color, and becomes zero in the colorless state? Fortunately, a theory with such a characteristic exists and is known as the Yang-Mills theory. It was invented by C.N. Yang, who already appeared in the parity non-conservation story, and R.L. Mills in 1954, and is known now by the general name of non-Abelian gauge theory; a detailed description of this theory will follow in a later chapter.

Are there Colored Hadrons?

Still, even if we have a convenient analogy between neutral atoms and colorless hadrons, this is not the last of the problems. Just as we can ionize atoms, we should be able to ionize hadrons and change them into colored states; we merely hypothesized that this ionization takes a lot of energy.

How much energy is this exactly and, what are the masses of the colored states? Since these states also include the individual quarks themselves, the question is closely related to the problem of the free quarks as well.

Let us now try to think of some tests to see if colored hadrons exist, or if indeed the three colored quarks exist. If hadrons can be ionized, then above some energy of collisions among hadrons, or hadrons and leptons, the ionization reactions should be possible; thus above some threshold, the collision cross section should increase and new resonances of colored states may be found. In other words, we expect to see a large increase in cross section at an energy very much larger than where we are now in hadron collision experiments.

To repeat: the quark is also one of the colored states, but an important distinction must be made here. Quarks have color and flavor, and the difference in flavor, u, d and s, also means a difference in charge. Thus among the colored states, there should be fractionally charged states such as quarks and also integer charged states which are like the ordinary baryons and mesons. In the former case, the direct measurement of charge is sufficient to identify the particles. The fact that the thickness of bubble chamber tracks is proportional to the square of the charge was mentioned before.

On the other hand, the integer charged states have the same electrical characteristics with or without color, so it is much more difficult to distinguish between colored and colorless states.

The Possibility of Integer Charged Quark

Perhaps the quarks themselves have integer charge and so are difficult to find. This possibility arises naturally once the three colors are introduced. The hypothesis here is that the charge not only depends on flavor but also on color. If a quark of a given color had an integer charge but the average over the three colors were fractional, the quarks inside colorless hadrons would look like the colorless fractionally charged quarks of Gell-Mann and Zweig.

This idea is called the Han-Nambu model. The figure shows the charge assignment explicitly.

According to this hypothesis, quarks are like normal baryons so the theory is close to the old Sakata model. But here quarks have much larger mass than the normal baryons and thus they may be unstable particles that decay into baryons and mesons. As yet there has been no evidence for such particles, be they fractionally charged or integrally charged, stable or unstable.

Color ＼ Flavor	u	d	s
Red	1	0	0
Green	1	0	0
Blue	0	-1	-1
Average	$\frac{2}{3}$	$-\frac{1}{3}$	$-\frac{1}{3}$

Fig. 11.2 Quark charge assignments in the Han-Nambu model.

But we do not have to give up yet. There is still one powerful test left. It is the reaction we spoke of earlier, of making hadrons from a collision of electrons and positrons. This reaction follows the steps shown below: (" " means virtual particles)

$$e^+ + e^- \rightarrow \text{``}\gamma\text{''} \rightarrow \text{``}q + \bar{q}\text{''} \rightarrow \text{hadrons.}$$

Thus the γ mediates the change of an electron pair into a quark pair, and then the quark pair, through the strong interactions, turns into several hadrons. Since the probability (cross section) of making a quark pair at a high enough energy is proportional to the square of its charge, the total cross section, which is obtained by summing over all reactions, should be proportional to the sum of all the quark charges squared. For example, for the Gell-Mann-Zweig u, d, s quarks, the sum of the squares of the charges is

$$R = (\tfrac{2}{3})^2 + (-\tfrac{1}{3})^2 + (-\tfrac{1}{3})^2 = \tfrac{2}{3}.$$

But if there are three colors, all 9 quarks must be added up; in this case $R = 2$. In the case of Han-Nambu quarks, $R = 4$, as can be seen readily from Fig. 11.2. (Here the total cross section includes the production of colored hadrons; if restricted only to colorless hadrons $R = 2$ as before.)

In the 1960's, the experimentally available energy was up to a few GeV and the value of R was measured to be close to 2, not $\frac{2}{3}$ or 4. This was thought to clearly indicate the existence of colored fractionally charged quarks. But what if the energy were raised higher?

One of the reasons the 8 GeV electron-positron collider called SPEAR was built at SLAC was to address the above question. Measurement of R is relatively simple since all the hadrons produced are counted regardless of their variety. When the experiment was carried out, the result was rather curious. It was found that the R becomes 4 above the energy of 4 GeV. This means that the Gell-Mann-Zweig three color quark model does not work. In the case of the Han-Nambu quarks, one can say that if only colorless hadrons are produced, $R = 2$, and if colored ones are included, $R = 4$; so perhaps colored hadrons are being produced at above 4 GeV.

The Unexpected New Particle J/Ψ

To clear up the above question, the people at SLAC carried out more precise experiments. In 1974, an unexpected occurrence took place: the discovery of the new particle J/ψ (J-Psi). It took quite some time before the nature of the J/ψ became clear. It was finally agreed that the J/ψ is not a colored hadron but a colorless hadron made of quarks with a new flavor called charm. In the next chapter, we will speak of J/ψ and other new particles.

12

CHARM AND ITS FOLLOWERS

The J/Ψ Indicates Profundity of Nature

The quark model was introduced to give an order to the hadrons which were discovered one after the other from the 1950's to the early 60's; according to the model, all hadrons are made from the three quarks u, d and s, and therefore, all objects in the universe are made of the three quarks and the four leptons (electron e, electron neutrino ν_e, muon μ, and muon neutrino ν_μ). If this were true, particle physics would seem to have arrived at a destination where the fundamental structure is only somewhat more complicated than what it was thought of at the beginning of the 20th century, when only the proton and the electron were known. What lies in the future would then be an investigation of the force field that mediates the weak interaction and one that mediates the strong force between the colored quarks.

The gauge field theory that describes the above force fields indeed made rapid progress in the 1970's; on the other hand, it was also found, experimentally, that the quarks and leptons had not been exhausted. The discovery of the J/ψ, mentioned briefly in the last chapter, was the first indication of it.

The J/ψ was discovered by S. Ting's group at MIT and B. Richter's group at SLAC independently and almost simultaneously. They jointly received the Nobel Prize in 1977 for this work. J was Ting's name and ψ was the SLAC group's name for the new particle. The experimental methods for the two experiments were completely different, but there was no doubt that the same particle was produced in both experiments. Ting's experiment used the AGS accelerator (30 GeV) at BNL (Brookhaven National Laboratory) to strike a Beryllium target with protons and looked for the e^+e^- decay of neutral hadrons produced in the pp (proton-proton) collision. The purpose of this experiment was essentially to look for new vector mesons.

Vector mesons are spin-1 mesons and, in terms of the quark model, correspond to a state where the spins of the q and $\bar{q}$ are parallel. In particular, the $u\bar{u}$, $d\bar{d}$, and $s\bar{s}$ pairs have charge zero and strangeness zero and thus are in the same class as the combinations $e\bar{e}$ (electron-positron) and $\mu\bar{\mu}$ (muon-antimuon); these mesons have decay modes, in addition to the decay into other hadrons; they also decay into leptons via the following reaction:

$$q + \bar{q} \to \text{``}\gamma\text{''} \to e + \bar{e}, \qquad \mu + \bar{\mu}.$$

The three vector mesons that were known to exist, ρ^0, ω^0, and ϕ^0, were thought to be appropriate combinations of the above three quark-antiquark pairs.

But are there other heavier vector mesons? Of course not all mesons are vector (spin-1) particles, but vector mesons are unique in that they always have the possibility of decaying into lepton pairs via a photon, which is also a vector particle; to look for vector mesons, then, one should look for high energy electron pairs. This was the idea behind Ting's experiment; the experiment was a large scale one, and the preparation took several years.

In the reaction

$$p + p(n) \to v^0 + x + y \ldots$$

the collision of the beam proton and the target proton (neutron) may produce a vector particle v^0 which may decay into an e^+ and an e^-. The mass of the parent particle is determined by the measurement of the energy of the latter particles.

Of course a quark-antiquark pair can decay into an electron pair even if the former is not a resonant state, so the electron energies are not necessarily fixed; if there is a resonance, however, there should be a sharp peak in the cross section at the resonant energy.

There was no guarantee that the new vector meson v^0 would exist; and even if it did, it would be produced with a small probability and the peak would be buried in the background and might not be seen. Ting's experiment was a kind of a gamble.

Richter's experiment was unique in that the equation on top is reversed; the experiment looks for hadrons being created in positron (e^+) electron (e^-) collisions. Since electrons do not interact strongly, the hadrons are made exclusively through electromagnetic interactions. Thus the above reaction is ideal for studying electromagnetic interactions, and the credit must go to

Richter who pushed the construction of the collider SPEAR. Even if there are no resonances, information concerning the number of quarks may be obtained, as was discussed earlier. Furthermore, according to the Weinberg-Salam theory, which is to be explained later, there exists a special vector particle called Z^0 (though not a hadron) that mediates the weak interaction, and it, too, will contribute to the electron-positron reaction. This is the reason why bigger and bigger electron-positron collider machines are being planned all over the world today.

The True Identity of the J/Ψ particle

Let us return now to the J/ψ particle. The reason the news of the discovery of the J/ψ was quickly reported in newspapers and became a sensation among the general public as well as among physicists must have been because of the dramatic nature of its appearance, especially as it appeared in the SLAC experiment.

The electron and positron beams go around the donut-shaped accelerator in opposite directions and cross at a certain point. The intersection is surrounded by detectors, which record

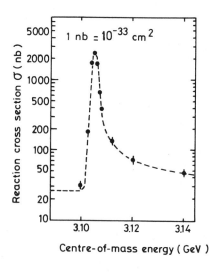

Fig. 12.1 The dramatic appearance of J/ψ.

the track of each particle that comes out; the characteristics and the energy of the particles are determined. Figure 12.1 shows the experimental data; as the energy of the beams is varied, the reactions occur at a slow rate at first but suddenly becomes an avalanche at 3.105 GeV. The cross section increases a hundred-fold within a narrow energy band of several million electron volts. Making an analogy with radios, this is like suddenly finding a setting for the radio tuner which gives a signal almost enough to break the speakers while all we found before were background noises.

SAMUEL C.C. TING   BURTON RICHTER

They are both American experimental physicists. They jointly received the Nobel Prize in 1977 for independently discovering the new particle J/Psi.

    - Ting is presently a professor at Massachusetts Institute of Technology, but the J particle experiment was carried out at Brookhaven National Laboratory. He has since participated in experiments at European facilities like DESY and CERN. The J particle seems to be named after the Chinese character (丁) that corresponds to his name.

Richter is one of the people who run SLAC (Stanford Linear Accelerator) and played an important role in the planning and construction of the electron collider facility SPEAR. This collider produced charmed particles such as ψ', χ(chi) and D after the discovery of the ψ particle. The new lepton τ (tau) was discovered at the same laboratory by M.L. Perl. SLAC had a monopoly of new particle discoveries for a few years.

The main features of a resonance are the height and width of the cross section. If the height of the resonance is high and the width narrow, the resonance stands out in dramatic relief; on the other hand, a low and wide resonance is difficult to see even if it is a resonance. If it is a resonance, the size of the cross-section, namely the area under the peak, is determined mainly by the energy and the characteristics of the reaction; this means that the height and the width are inversely proportional to each other. A narrower peak is that much taller.

The known vector mesons, that is to say, the ρ^0, ω^0 and ϕ^0, indeed show up as resonances, but their widths are as large as 100 MeV. Thus the discovery of a sharp peak such as that of the J/ψ had a much larger impact on physicists than having simply found another meson.

In general, the width of a resonance is inversely proportional to its lifetime, according to the uncertainty principle. In other words, the narrower the width, the longer the lifetime, and the more nearly stable the resonance is. What exactly is the J/ψ which is much more stable than other mesons?

It took two years before the true identity of the J/ψ became clear through further experimental and theoretical work. Starting with the experimental data, first a similar peak (called ψ') was discovered at an energy just above that of the J/ψ about a week after the discovery of the ψ. At even higher energies, resonances which are wider and more like normal mesons were found around 4 GeV. As we mentioned before, the total cross-section for $e^+ + e^- \rightarrow$ hadrons rises in general even where there are no resonances (the R gets larger). The actual data is shown in Fig. 12.2.

All the resonances mentioned above have the same spin, parity, and other quantum numbers as the photon, but a different group of mesons called χ (chi) were also discovered. Chi particles are made when a ψ' gives off a photon. Thus:

$$\psi' \rightarrow \chi + \gamma.$$

The masses of these particles are between those of the ψ and the ψ'. Their spin and parity are believed to be 0^+. Another particle, the more recently discovered η_c (eta c), has a mass just slightly smaller than that of ψ and is thought to have spin-parity 0^- like the π^0 and η^0.

When the mass spectrum of these mesons is examined, the cluster of masses at around 3-4 GeV reminiscent of, say, the spectrum of the hydrogen atom. According to the now prevalent view, these resonances correspond to the various levels of the compound state of the fourth quark c (charm) and its antiparticle $\bar{c}$; this is exactly like the well known positronium (the bound state of an electron and a positron). But the binding force is not Coulombic but is based on color.

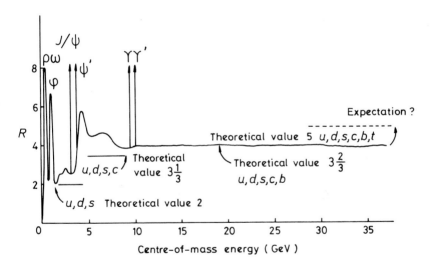

Fig. 12.2 The changing values of R. Note the prominent resonances at 3 and 9 GeV.

The Fourth Quark c that Completed the Quark Model

The above interpretation of the J/ψ is based on the existence of the c quark and is difficult to justify from the data on the psi and the chi particles alone. The conclusive evidence came from the discovery of the D and F-mesons in 1976. These mesons have mass about half of that of the psi particles and are thought to be the compound state of a c quark and one of the normal quarks u, d or s.

$$D^+ = c\bar{d}, \qquad D^0 = c\bar{u},$$
$$\bar{D}^0 = u\bar{c}, \qquad D^- = d\bar{c},$$
$$F^+ = c\bar{s}, \qquad F^- = s\bar{c}.$$

As can be seen from above, the c quark is given the electric charge $\frac{2}{3}$, the same as that of the u quark. The spin-parity of the D and F-mesons are the same as that of the η^-, namely 0^-, which means that these mesons are in the same family as the pions, K-mesons, and eta (η) mesons. The D*'s and F*'s, which are slightly heavier than the D and F-mesons and are the members of the same family as the rho (ρ), omega (ω), K*, and psi (ψ) vector mesons of spin-parity 1^-, are also known.

D-mesons and K-mesons are quite similar; D-mesons can be made by replacing the s quark in the K-meson with the c quark. Just as the s quark carried the strangeness of the K-meson, the D-meson with a c quark carries the new quantum number "charm". So the F-meson carries charm and strangeness at the same time.

The quantum numbers charm and strange are useful because they are almost conserved; the flavor changes only through weak interactions. Thus D and F-mesons are relatively stable and their decays are thought to take place through the decay of c quark into s quark:

$$c \rightarrow s + \nu + \bar{e}, \quad s + u + \bar{d}.$$

Turning to the mass of the c quark, the mass of the hadron is more or less the sum of the masses of the quarks. According to this assumption

$$D \sim 1.8 \text{ GeV}, \quad u, d \sim 0.3 \text{ GeV}.$$

so that

$$c \sim 1.8 - 0.3 = 1.5 \text{ GeV}$$

which agrees with the mass of the psi

$$\psi = c\bar{c} \sim 3 \text{ GeV}.$$

The consistency of the interpretation makes the acceptance of the charm quark theory natural; however, the concept of charm originated a long time before the discovery of the ψ and was conceived as a theoretical tool to understand the weak interactions. I will leave the detailed explanation of this to a later chapter and just state the essence of the idea. The normal beta decays such as:

$$u \rightarrow d + \nu_e + \bar{e}$$
$$\mu \rightarrow \nu_\mu + \nu_e + \bar{e}$$

occur as a change between pairs of quarks or leptons such as (u, d), (μ, ν_μ), or (e, ν_e). So both quarks and leptons must come in even numbers of types (flavors) which form pairs. But then the s quark is left by itself so its partner quark c must be hypothesized. Then the pair (c, s) exists in correspondence with the (u, d) group and the charges ought to be ($\frac{2}{3}$, $-\frac{1}{3}$) for both.

This was a rather daring hypothesis, but it has a convincing simplicity. In reality, however, s quarks can also decay into d quarks to cause V particle decays; the weak interaction is quite complex and mysterious, and it took a long time for this subject to be completely understood. Even experimentally, it took ten years for the c quark to be discovered.

Outdone by Nature Again

Are we done with everything now that the quark model is theoretically complete with the discovery of the c quarks? If so, the four kinds of leptons (ν, e), (ν_μ, μ) and the four flavors times three colors of quarks (u_i, d_i), (c_i, s_i), (i = red, green, blue) should be all of the elementary particles.

But nature again proved our simplistic expectations false and taught us that other kinds of elementary particles also exist.

The first of these surprises was the discovery of the new lepton τ (tau) at SLAC by M.L. Perl's group. The experiment again used the electron-positron collision at SPEAR; they found the particle among the collision products, which is a lepton like the muon but of much higher mass of 1.8 GeV. The decay of this particle goes as:

$$\tau \to \left\{ \begin{array}{l} \nu_\tau + \mu + \bar{\nu}_\mu \\ \nu_\tau + e + \bar{\nu}_e \end{array} \right.$$

and corresponds exactly to the decay of the muon. The nature of the new neutrino ν_τ is not entirely clear, but there is no doubt that the pair (τ, ν_τ) exists.

Since there are now six kinds of leptons, there might be six kinds of quarks. As if to answer such anticipations, L. Lederman's group in 1977 discovered a meson at energies above the J/ψ and called it Υ (upsilon). It was a proton-proton collision experiment, as in the case of Ting, but the Fermilab (Fermi National Accelerator Laboratory) 400 GeV proton accelerator was used. Later the result was confirmed also in the electron-position reactions at Hamburg and Cornell University; these new experiments also found the excited states of the Υ, designated Υ' and Υ''.

Since the upsilon family has masses around 10 GeV, the supposed fifth quark in the upsilon must have a mass of about 5 GeV. Deducing from the patterns indicated by the known pairs (u, d) and (c, s), the new quark must correspond to the member b of the next pair (t, b), the member whose charge is $-\frac{1}{3}$. t and b are usually called top and bottom; however, the more aesthetic names of truth and beauty are also used.

The top has not yet been found. We know, on the other hand, that the ϕ^0, which carries $s\bar{s}$, has a mass around 1 GeV, the J/ψ, which is $c\bar{c}$, has a mass around 3 GeV, and the γ, which is $b\bar{b}$, has a mass around 9 GeV; so some people speculated that the mass of the $t\bar{t}$ may be around 27 GeV. Unfortunately, the new electron-positron machine PETRA at Hamburg, whose energy went up to over 40 GeV, turned up no such particle, thus disappointing the German scientists who wanted to scoop the Americans.

I spoke earlier about the fact that the cross section of the reaction electron-positron $\to$ hadrons is a measure of the number of quarks. Let me here repeat the explanation. One can get a rough estimate of the total cross section of the reaction $e^+ + e^- \to$ hadrons at a given energy by computing the sum of the squares of the charges of the quarks that can be produced at that energy. This is the quantity called R (p. 120). Below, R is computed for the energies below 3 GeV (where only u, d, and s quarks can be created), between 3 and 9 GeV (only u, d, s, c), and above 9 GeV (u, d, s, c, b).

$$\left\{ \left(\tfrac{2}{3}\right)^2 + \left(\tfrac{1}{3}\right)^2 + \left(\tfrac{1}{3}\right)^2 \right\} \times 3 = 2$$

$$\left\{ \left(\tfrac{2}{3}\right)^2 + \left(\tfrac{1}{3}\right)^2 + \left(\tfrac{2}{3}\right)^2 + \left(\tfrac{1}{3}\right)^2 \right\} \times 3 = 3 + \tfrac{1}{3}$$

$$\left\{ \left(\tfrac{2}{3}\right)^2 + \left(\tfrac{1}{3}\right)^2 + \left(\tfrac{2}{3}\right)^2 + \left(\tfrac{1}{3}\right)^2 + \left(\tfrac{1}{3}\right)^2 \right\} \times 3 = 3 + \tfrac{2}{3}$$

The experimental values are shown on Fig. 12.2. It can be seen that the R value, except for the the psi and upsilon resonances around 3 and 9 GeV, agrees with the theoretical values above.

At Least 6 Leptons and at Least 5 Quarks

Let us now summarize what we learned in this chapter. The number of the kinds of leptons and quarks have increased since the introduction of the quark model in the 1960's, and six leptons and at least five quarks are presently known. We still lack reliable theories to tell us how many elementary particles there are and what their masses should be. We may say that we have no ways to find out other than building bigger accelerators and looking for new particles.

But it is not true at all to say that the theories are useless. The existence of the charmed quarks was predicted long before the discovery of the J/ψ particle. The invisible quantum number, color, came about from purely theoretical needs; even the existence of quarks themselves would become suspect without the theories. The planning of larger and larger accelerators by physicists is not based on an irresponsibly vague hope of finding something new.

The progress of theories since the 1960's, in parallel with the discovery of the new particles, has been remarkable. Many new concepts were introduced and their effectiveness proven. Though still incomplete, theories that at once describe all the elementary particles and all the force fields can now be thought of as a real possibility rather than a dream. And according to such theories, the particles and fields have not yet been exhausted.

13

QUARKS WITH STRINGS ATTACHED

A Paradox

Let us now go to an earlier point in the story. The quark model not only explained the SU_3 symmetry very well but could also account for the various spin states of the hadrons with some degree of success by invoking the SU_6 theory. Here the quarks act as though they were weakly bound in hadrons. This is clearly a paradox because quarks are never seen to come out of the hadron.

Hadrons also have other important characteristics not yet mentioned here which one would like to understand in terms of the quark model. To this end, one must be prepared, if necessary, to add new elements or assumptions to the model. The purpose of this chapter is to present the relevent experimental facts and the models, or theories, that are designed to explain these facts.

Hadron collisions, in general, result not only in elastic scattering but also in the creation of several hadrons. As it often happens, the cross section for the reaction varies rapidly with energy and thereby signals the presence of resonances, or the excited states of the hadrons. Examination of the energy distribution of the produced hadrons will further reveal that, a small number of excited states are first created and then decay into several hadrons.

Let us consider two questions here. First, how many excited states are there to a hadron? Are there an infinite number of resonances as the energy is raised indefinitely? Secondly, how does the reaction cross section vary as the energy is raised? And how about the number of particles created and their energy distributions?

First let us consider the answers to the above questions in a pictorial form. The relevant figure is Fig. 7.7 on p. 75 in which the variation of the total cross section for $\pi^+ + p$ with energy is plotted. Thus the relevant process is:

$$\pi^+ + p \rightarrow x$$

where x includes all possible reactions. What we learn from this figure is:

1. Resonances are repeated many times but become less prominent at higher energies.

2. The cross section seems to gradually decrease but also seems to be approaching a constant limit.

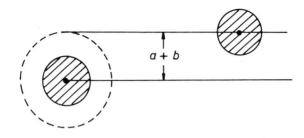

Fig. 13.1   A ball of radius b striking a bigger ball of radius a.

A comment on the second point first. The fact that the cross section reaches a finite value may be interpreted as the result of the hadron possessing a fixed size. The cross section σ for striking a target of radius b with a ball of radius a is

$$\sigma = \pi (a + b)^2$$

since the center of the ball must be within the distance $a + b$ from the center of the target. In the case of the elementary particles, the cross section can be even larger than this because of the wave mechanical spread; however, at high energies, that is to say for short wavelengths, the above classical interpretation is roughly correct. So the hadron radius can be estimated from the size of σ to be about 1 fermi (10^{-13} cm), but of course this was already mentioned.

Next we turn to the resonances. The fact that the cross section curve becomes smoother with rising energy does not necessarily indicate that resonances are disappearing. Since many reactions are possible at high energies, even if a resonance occurred in one of the reactions (in a given angular momentum state) it would constitute only a small portion of the total cross section and may be invisible under the background; and if wide resonances occurred one after the other in rapid sucession, the cross section may be averaged out.

Aside from these points, we may also ask what kind of regularity is exhibited by the resonances. The first things we are interested in are the mass and spin of a resonance. The spin, or the angular momentum of the resonant state, is determined by analyzing the angular distribution of the scattering. As stated before, the complexity of the dependence of the scattering on angle increases with the angular momentum.

The graph that plots the relationship between the square of the mass (rest energy) and the spin of resonances is called Regge trajectory. This term comes from the theory constructed by the Italian theorist T. Regge. The trajectory is a beautiful straight line in the case of the $\pi^+ + p$ (isospin = $\frac{3}{2}$) reaction as may be seen in Fig. 13.2. The resonance located at spin $j = \frac{3}{2}$ on the trajectory is the familiar 3–3 resonance; j then increases in units of two to $\frac{7}{2}$, $\frac{11}{2}$, . . . and so on.

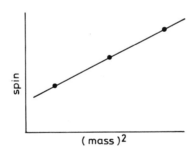

Fig. 13.2 A Regge trajectory.

The result is more or less the same for other reactions such as the $\pi^- + p$, $K^- + p$. These correspond to the Regge trajectories of the baryon resonances that differ in the internal quantum numbers (strangeness, isospin, and so forth). The same kind of analysis should also apply to the meson case of meson-meson reactions like $\pi + \pi$; however, this experiment cannot be done in reality since the pions themselves are unstable. Nevertheless, the existence of mesonic resonances with spin 0, 1, . . . are known from their creation and decay (for example, p + p → ρ + . . . , ρ → π + π); when these are plotted as a Regge trajectory, they again form a straight line.

These various Regge trajectories are not only straight but are parallel to each other. Thus the rate of the increase of the (square of) mass with spin is the same for all resonances; only the starting point is different due to the difference in the internal quantum numbers. This regularity of the Regge trajectories is a surprising fact and must be a result of some deeper principle. If the straight line is to continue forever, the resonant states must recur up to infinite energy and infinite spin.

From the point of view of the quark model, it would seem that the fact that there are resonances at arbitrarily high energy indicates that the quarks can never become free. But a more detailed explanation is needed.

String Model of Hadrons

Now moving on to the theoretical explanation of the Regge theory; it is easier to start with the result than to trace the actual development of the ideas. The result is what is called the string model of hadrons According to this model, hadrons are quarks bound to each other by something like rubber bands. Figure 13.3 shows the structure of a meson and a baryon; there is a quark at the end of each string. The difference between a string and a rubber band is that the string can stretch indefinitely while the tension remains the same, somewhat like a spider's thread. It is

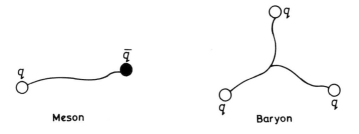

Fig. 13.3   The string model for hadrons.

also possible for a string to break; when this happens, new quarks form at the broken ends of the strings.

Now consider the internal motion of mesons. It is possible to have a yo-yo-like motion where the string stretches and contracts; it is also possible to have a rotating motion like a hammer throw. In the latter case, the centrifugal force and tension balance each other out and the system looks no different from a rotating stick. The equation to describe the balance can be easily written down if the energy equals mass principle is applied to the potential energy of the tension stored at each point in the string. The length of the string is governed by the fact that the velocities of the ends of the string cannot exceed the speed of light; this leads to the peculiar conclusion that the slower the rotation, the longer the string. At any rate one gets the correct relationship between the total energy E of the string and the angular momentum l, namely, $l \propto E^2$. The constant of proportionality in this relation is determined by the tension so the conclusion is that the Regge trajectories have a universal slope.

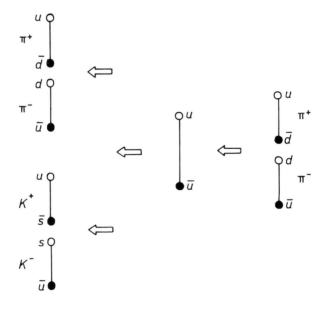

Fig. 13.4 The breaking and joining of two strings.

The argument for baryons is not quite so simple. A question may arise in the reader's mind as to why the strings should be attached to the three quarks in the way shown on Fig. 13.3. The reason comes from the necessity to relate the concept of string to the gauge field of color, but it will not be discussed further here.

The process of cutting and joining strings must also be explained; this is quite important in the description of the hadron reactions. For example, think now of the case of $\pi^+ \pi^-$ collision. Here the ends of the $\pi^+ = u\bar{d}$ string and the $\pi^- = d\bar{u}$ string may come together resulting in annihilation of a $u\bar{u}$ pair or a $d\bar{d}$ pair and formation of a temporary state of a single string which subsequently breaks again into two mesons. The meson end products may not necessarily be $\pi^+ \pi^-$ but $K^+ = u\bar{s}$ and $K^- = s\bar{u}$, or $K^0 = d\bar{s}$ and $\bar{K}^0 = s\bar{d}$.

In Fig. 13.5a the motion of the strings in the above process is followed as time goes on. One string flows in time making a

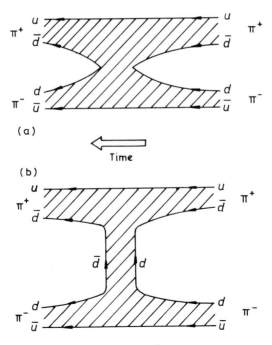

Fig. 13.5   The exchange of a $d\bar{d}$ meson by π^+ and π^-.

ribbon; the edges of the ribbon are the quarks. For a while the two ribbons become one, like Siamese twins.

A gradual deformation of Fig. 13.5a will result in the Fig. 13.6b. This figure represents nothing other than an exchange of a kaon between two mesons, a Yukawa-type process. If this last figure is further rotated by 90 degrees, it becomes that of the scattering between a K^+-meson and a π^--meson (Fig. 13.7). In this case the particle with the arrow pointing in the opposite direction from the previous case must be reinterpreted as an antiparticle.

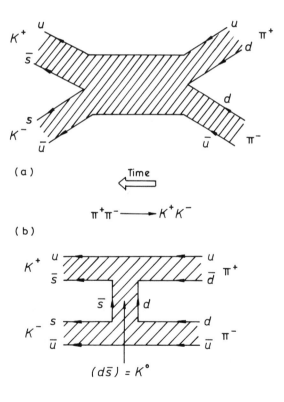

Fig. 13.6   The exchange of a K-meson by two mesons.

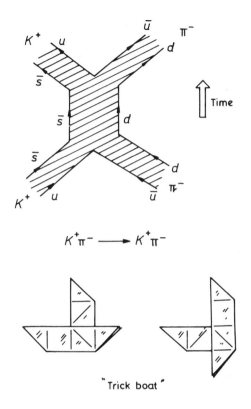

Fig. 13.7   Hadron scattering is like a "trick boat".

When I was a child, I liked an origami called "trick boat". As the readers may be aware, this origami is of a sailing boat whose sail can become the bow in an instant; hadron scattering is something like this origami. Technically this phenomenon is known, by the impressive name of "duality principle".

Let us now try to use the string model in the electron-positron reaction. The quark pair created by the reaction $e^+e^- \to \gamma \to q\bar{q}$ will be shot out in opposite directions at high energy. Since there is a string between the q and $\bar{q}$, they must eventually come back together if the string does not break. It is, however, possible that the string may break up into several pieces; then each of these pieces will come out as a meson. Try making a string from chewing gum and stretching it suddenly; it is difficult to break it into three pieces, but in the case of the hadron, the number of pieces, i.e. multiplicity, will increase with the energy.

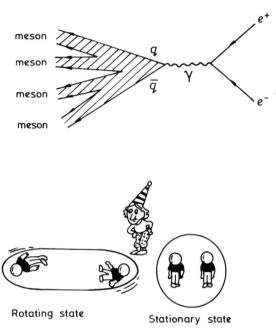

Fig. 13.8   An electron-positron reaction.

What is A String?

What is a string exactly? As will be explained in more detail, the string is presently believed to be something like a bundle of electric field lines. But for the sake of analogy, it is more convenient to think of magnetic field lines. The string corresponds to a bar magnet and quarks to its poles. A bar magnet cut in two becomes two magnets; it is impossible to isolate a north or south pole by itself. A single quark, it appears, does not exist.

If there is a meaning to the above analogy, the string should not be thought of as a mathematical line but as an object having some thickness. An especially short string, a low energy state, should have about the same length and thickness. K. Johnson's bag model was introduced to describe such a state. According to this model, a hadron is something like quarks trapped in a rubber balloon. If this balloon is rotated rapidly, it will get elongated and look like an airship; this is nothing other than the string.

14

WHAT IS A PARTON?

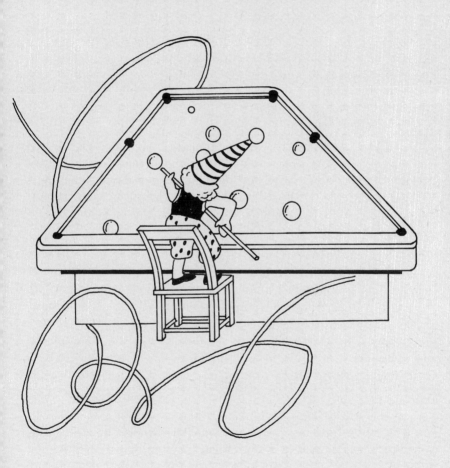

Hadrons are Soft

Hadrons have a complex nature; the string model does not take care of every contingency. In fact hadrons have characteristics that completely contradict the predictions of the string model. In this chapter we will talk about the other side of the split-personality hadron.

We spoke before about the fact that the resonances disappear at higher and higher energies in hadron-hadron collisions, and that the cross section becomes smooth. It is this rather boring region, in which the cross section is smooth, that we now want to discuss. Successive experiments made clear that the characteristics of the hadron in this energy region are just about as boring as they can possibly get.

The main reaction that dominates high energies is the multiple particle production. For example, the average proton-proton reaction at CERN or Fermilab creates about 10 hadrons; the majority of these are pions. In this case, it is easier experimentally, and for analysis also, to treat the created particles statistically and look at, say, their energy distribution.

Let me now explain what an actual reaction looks like. If two protons going in opposite directions collide, these break into many fragments (hadrons); these fragments seldom fly away in the sideways direction but appear as jets in the forward and backward directions. The width of the jets is inversely proportional to the initial energy. All this is proof that hadrons are quite soft.

Recall Rutherford's hypothesis which established the structure of the atom. Rutherford hypothesized that the reason the angle of scattering is large when alpha particles and atoms collide is because there is a heavy and small nucleus in the atom. Our experimental situation is the complete opposite; since hadrons do not have hard cores, the motion is not impeded too much in collisions giving rise to forward jets.

What happened to the quarks in hadrons? Can't quarks collide with each other just as in the Rutherford scattering? The string model is still necessary to explain the jets in terms of the mechanism of a string breaking into many pieces.

Some data that shed light on the above problem came from SLAC around 1967-8. These data did not come from hadron-hadron scattering experiments but from the electron-proton reaction experiments using the 20 GeV electron linear accelerator. In fact, the latter reaction is more closely analogous to the Rutherford experiment than the former. Instead of striking the large atom with the point-like alpha particle (helium nucleus), the large proton is struck with the point-like electron. (Of course, the size scale is completely different between the two cases.) Furthermore, the electrons do not interact strongly, and their electromagnetic characteristics are well known; this is convenient for analysis.

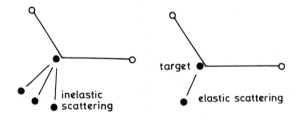

Fig. 14.1 Elastic and inelastic collisions between particles.

R. Hofstadter had been doing this kind of experiment at Stanford for quite a long time in order to study the charge distribution of nuclei; he won the Nobel Prize for his work. He studied, however, the elastic collisions (the internal structure of the particle remains the same after the collision, and only the momentum changes); in the present case, we look for inelastic collisions, that is to say, collisions in which many hadrons are created. The quantities measured in the experiments are the energy and momentum of the electron after the collision. What happens to the target proton is not detected. In the case of elastic collisions, the energy of the electron is determined immediately by the scattering angle. If proton breaks up in the reaction extra energy is spent so the electron scattered at a given angle should have less energy than in the elastic scattering case. Thus, the energy distribution of the electron that is scattered in a particular direction reflects how the proton broke up.

Infinitely Small Point Particle

As a result of these experiments, a simple law called scaling was found. Simply put, this law describes the fact that the measurement of the energy and angular distribution of the scattered electron at a certain initial energy only differ from the result at another intial energy by some constant. In other words, if the scale is changed appropriately for each energy, all the data will fall on the same curve. This is shown in Fig. 14.2; the x on the horizontal scale reflects the elasticity of the collision. $x = 1$ means elastic, and $x = 0$ is completely inelastic, meaning the reaction in which the electron sticks to the proton.

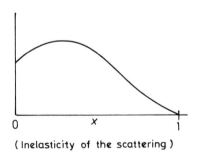

(Inelasticity of the scattering)

Fig. 14.2 Bjorken scaling.

Since this scale law was predicted by the theorist J. D. Bjorken before the experiment, it is generally called Bjorken Scaling. The meaning of this scaling law is that there is no set scale inside the hadron. Again taking the example of the atom, the atom has two scales; the size of the entire atom and the size of the nucleus. In quantum mechanics, the length scale is the inverse of the momentum scale; this means that if there are two scales, the phenomena may differ fundamentally at low energies and at high energies. If there is only one scale, however, there should be no new phenomena no matter how high the energy beomes. In this case, if there is a next scale, its size could be considered infinite. Thus, if there are constituent particles in hadrons, they must be infinitely small point particles.

Feynman's Parton Model

The model that makes the above ideas quantitative was introduced by R. P. Feynman. Feynman is well known as the originator of the Feynman diagrams; his theory here stated that the hadrons are made of many particles. He called these particles partons (i.e. partial particles).

The description of the electron-proton reaction according to the parton model goes something like this. The proton behaves like a gas of partons, and the reaction is a process of collision between the electron and one of the partons. Partons do not have a fixed mass; partons of all masses exist in the proton in ratio.

RICHARD P. FEYNMAN

He was born in New York in 1918 and took his degree at Princeton in 1942. He was involved in the Manhattan Project during the war. In physics, he is know for developing unconventional points of view; he formulated quantum mechanics with a method called Feynman's path integrals. Today, his method has become the orthodox way. As a natural extension of this method, he originated the interpretation that antiparticles are particles that move backwards in time; he also invented one of today's standard calculational methods called Feynman diagrams. He received the Nobel Prize in 1965 along with Tomonaga and Schwinger.
Other famous works include the Feynman-Gell-Mann V minus A theory (1957) and the parton model (1969).

Feynman is an irreverent man with a rich sense of American humor. At conferences, he becomes a star who captivates his audience. His textbook, written for university students, is very famous, but it is quite a task to completely digest the contents.

The number of partons is not fixed either; the only requirement is that the sum of the energies of all the partons should equal the total energy of the proton. When an electron and a proton collide, the electron collides with one of the partons and the latter gets knocked out; the relationship between the scattering angle and energy is determined by the masses of the electron and the colliding parton. (This is something like playing billiards with balls of different masses.) Thus, the curve on Fig. 14.2 is thought to reflect the parton energy (mass) distribution. The large x region represents collisions with heavy partons, the small x region those with light partons; according to the figure, there are relatively more of the lighter partons than of the heavier ones in the proton.

The strength of this rather unusual parton model lies not only in the fact that the scaling law can be explained but also in explaining the experimental data in terms of the parton probability distribution. Having fixed the distribution, this can be used not only for the electron-proton scattering but for, say, the proton-proton scattering. In the latter case, the scattering between two partons is considered; a kind of scaling law is applicable here also. Here, however, the scattering is not only electromagnetic but also depends on the strong force. What are measured are the energy distributions of the created hadrons. The fact that many low energy hadrons are created from light partons is similar to the electron-proton reaction case.

Comparison with the Quark Model

The ideas in the parton model have certain similarities with the hadron models we spoke of earlier, but there are also points which are mutually contradictory. It may seem natural to say parton equals quark, but there are immediate problems; there is no set number of partons, for example. The proton does not consist only of three partons but also contains many light partons. Partons, by hypothesis, do not interact with each other, but the quarks are supposed to be tied with strings.

There is one other problem with the electron-proton scattering data. They imply that there are partons which are neutral, not carrying any charge. Neutral partons cannot collide with an electron but carry a part of the total energy of the proton. So the total energy observed through the electron scattering does not tally with the proton's total energy. According to actual data, it must be interpreted that about half of the energy of the proton, on the average, is carried away by neutral partons.

These kinds of contradictions between the two models is a reflection of the complex characteristics of the hadron itself. It must be thought that each of the models merely reflects a portion of the reality of the hadron. As Sakata emphasized, models are not final theories. We must hope for the appearance of more fundamental theories.

Fortunately, this fundamental stage seems to already have arrived. The theory concerned is the QCD (quantum theory of the color gauge field), which will be explained in a later chapter. Let me here just briefly explain how the above inconsistencies are resolved in this theory.

Faraway-string, Close together-gluon

The string describes the behavior of the force at long distances, the parton, at short distances.

The neutral parton is the quantum of the gauge field of color, that is to say, the gluon. The interquark force, according to the gauge field of color, differs from the electromagnetic force in that it becomes stronger at longer distances and weaker at shorter distances. The string describes the behavior of the force at long distances, the parton, the behavior at short distances. The reason there are no fixed number of partons is that gluons are continually transforming themselves into quark-antiquark pairs and vice versa.

15

TOMONAGA'S RENORMALIZATION THEORY

Particle Physics is Busy Chasing New Phenomena

Readers who have gotten this far in this book may have gotten the impression that particle physics is merely a collection of models which are able to qualitatively explain various phenomena but cannot supply quantitatively accurate answers, and that there is no fundamental theory that can summarize everything.

It is difficult to respond to such a criticism. Compared to Newtonian mechanics which can predict the time and place of an eclipse, or Einstein's general theory of relativity which predicts that the movement of the perihelion of the planet Mercury is 40 seconds per 100 years, elementary particle physics, in general, is very much poorer in predictive power; indeed, the name "exact science" may not be applicable here. In elementary particle physics, we are kept so busy chasing after new phenomena and new realities that the problem of qualitative understanding must come first.

Quantum Electrodynamics

There are, however, areas of particle physics in which our predictive ability has reached a region of very high sensitivity. One of these areas is called QED (quantum electrodynamics) and it concerns the electromagnetic characteristics of particles. Especially the study of electromagnetic characteristics of leptons, such as the electron and the muon, has achieved a very high precision both theoretically and experimentally; futhermore, theoretical and experimental values match exactly.

The magnetic moment of the electron is a typical example of such characteristics. The value of this quantity deviates by a very small amount from the naive value obtained from the Dirac equation, which is equal to twice the value of the Bohr magneton. The correction to it is expressed by a multiplicative constant:

experimental value : 1.001159652200
theoretical value   : 1.001159652415

The skills that are required to achieve such a high precision in experiment are enormous; the theoretical calculations at this level are also quite sophisticated. According to Toichiro Kinoshita (Cornell University), an authority in this field, his calculations that attempt to raise the precision of these figures take several hundred hours even using a giant computer. The agreement between theory and experiment is one of the successes of QED.

The reason people do not hesitate to undertake such enormous efforts is because QED is trusted completely. Saying it in a different way, if an inconsistency occurs somewhere, that in itself would be a big discovery. (The small difference in the magnetic moment value above is not thought to be a real effect.) This QED theory was completed in the 1940's by S. Tomonaga, J. Schwinger, R.P. Feynman, F. Dyson and others, and is an exemplary theoretical structure to theorists who tackle elementary particle physics.

Infinite Self-energy

The true nature of QED resides in the concept of "renormalization". Tomonaga called this concept by a name, in Japanese, which connotes "compounding of interest" in bank accounts, that is to say, putting the interest back into the account to earn more interest. The latter term seems, to me, to be more descriptive of the actual theoretical process.

In any case, QED treats the interaction between electrons and the electromagnetic field quantum mechanically; for electrons, the Dirac equation becomes the basic equation; for the electromagnetic field, Maxwell's equations play the same role. One of the manifestations of quantum theoretical effects is that electrons are releasing and capturing virtual photons, and the photons are turning into virtual electron-positron pairs and back into photons again all the time. Electromagnetic forces are created as a result of exchange of photons between two electrons; it is also possible for one electron to emit and reabsorb photons. This latter is a phenomenon in which the charge of the electron interacts with itself; as a result, the electron acquires a kind of potential energy called electromagnetic self-energy. The mass of the electron is assumed to be the sum of the mechanical mass and the electromagnetic self-mass.

The above problem appeared in classical electromagnetism as well, and, a long time ago, H.A. Lorentz took this problem up in detail in his famous electron theory. If a metal sphere of radius r is given a charge e, the self-energy in the order of e^2/r appears because of the Coulomb force between its elements. If then, the radius r is made smaller, the self-energy becomes larger without limit. Since the self-energy of a sphere of radius about 1 fermi (10^{-15} meter) is about the same as the rest energy of an electron, Lorentz theorized that 1 fermi is about the size of the electron and its mass is entirely made up of the electromagnetic self-energy.

But if Lorentz was right, the electron must have a structure and is not a point particle. If there is a true elementary particle, it must be a point particle, which means that the infinite self-energy cannot be avoided after all.

It was discovered by V. Weisskopf in the 1930's that the conclusions based on quantum field theory are slightly different

---

SIN-ITIRO TOMONAGA (1906-1979)

He was born in Tokyo and grew up in Kyoto. His father was the philosopher Sanjuro Tomonaga. Both his age and environment were close to those of Hideki Yukawa; both men chose particle theory as a career and became rivals for life. Unlike Yukawa, he worked in the Nishina Laboratory which was a part of the Institute for Physical and Chemical Research (Riken) in Tokyo; he later studied in Germany under W. Heisenberg.

During the war (1943), he began what is called super-many time theory which developed until it eventually led to the completion of renormalization theory, the explanation of the Lamb shift and so on. He received the Nobel Prize in 1965 along with Schwinger and Feynman. He also worked on cosmic ray physics in partnership with Y. Nishina. During the war, he worked on theories of magnetrons and microwave circuits (wave guides), among others. He was also a fine teacher and guided many students. He possessed fine Japanese sensibilities, and the year he spent in U.S.A. (1949 in Princeton) seems to have been a hardship. "I have been exiled to heaven," he complained to his students.

from those of the classical theory. This difference is due to the fact that the electron has a cloud of virtual electron-positron pairs around itself, and therefore the charge is, in reality, dispersed over an area (Fig. 15.1). This dispersion does not make the self-energy finite, but the degree of infinitude becomes logarithmic, which is rather less extreme than before.

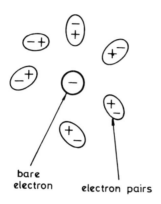

Fig. 15.1 Virtual electron pairs surrounding the point electron.

Applying Lorentz's argument, the size of the electron now becomes 10^{-30} cm instead of 10^{-13} cm. We can't simply say, however, "let's forget about it" since the present state of elementary physics cannot reach such a small domain. This is a matter of fundamental principles. Furthermore, we get infinite numbers if we try to push through calculations of electron scattering cross sections and other observable quantities; we might be forced to conclude that quantum electrodynamics is nonsense.

Renormalization Theory — An Act of Giving Up Something

What saved QED by conquering the above difficulty was the theory by S. Tomonaga, J. Schwinger, R.P. Feynman, and F. Dyson. To explain the idea behind this theory, it seems appropriate to quote the phrase "principle of renunciation" used by

Tomonaga from time to time. The meaning of this phrase is that one should give up the hope that the theory is perfect and that everything can be calculated from it, and, instead, one should make a definite distinction between things that can be calculated and those that cannot; in some ways, it smacks of Eastern philosophies.

· The reason this kind of philosophy succeeded as physics is because Tomonaga and others showed that the division, between things that can be calculated and those that cannot, could be carried out rigorously in the mathematical sense. In the present case, those things that cannot be calculated are the mass of the electron (including the self-energy) and the charge of the electron; the former was explained earlier. The meaning of the latter indetermination is that, referring to Fig. 15.1, the electron-positron pairs around the electron interacts with the initial charge with the result that the electrons are pushed away and the positrons are pulled in; thus, the initial point charge becomes covered with a cloud and the measured net charge becomes less. The calculation of this effect also leads to the nonsensical answer of infinity.

Tomonaga and others pointed out that what are actually measured experimentally are the total mass and total charge which cannot be divided into the bare part and the cloud part. Thus, even if the cloud part is infinite, the problem of the infinity is solved if the total in which the cloud part is renormalized into the bare part is finite, and represents observed mass or charge. In a figurative sense, the capital is an infinite profit, and the compounded interest is an infinite loss, but the final sum is finite.

Perhaps it cannot be denied that the theory of Tomonaga and others treats only the symptoms of the problems in QED. But as the result of this theory, an answer can be found to any electromagnetic process and the accuracy of the calculation can be improved at will. The theory made perfect quantitative predictions of phenomena such as the magnetic moment of the electron, mentioned above, and the famous Lamb shift. The figure in the frontispiece gives a simile of the state of affairs arrived at by QED.

One of the reasons for the success of QED is that the electric charge e happens to be a relatively small number. In QED, the parameter that expresses the strength of electromagnetism is Sommerfeld's fine structure constant e^2/hc $= \frac{1}{137}$. Since the probability of each emission or absorption of a photon is proportional to this constant, the processes in which many photons are involved are suppressed. Thus, the method of perturbation theory in the powers of $\frac{1}{137}$ is effective here. (Kinoshita's calculation, mentioned earlier, of the anomalous magnetic moment corresponds to the fourth power of $\frac{1}{137}$.)

Bare charge cannot be seen because of the intervening virtual pairs.

However, we learned already that electromagnetism is not the only interaction that exists in nature. It would be natural, therefore, to apply the successful methods of QED to other interactions; this is easy to say but very hard to actually carry out.

The reason for this difficulty is that both strong and weak interactions are quite complex and their true nature is not well understood. Both forces are short ranged, which is different from the case of electromagnetism; the renormalization theory is, in general, ineffective here. And in the case of strong interactions, the calculational method of QED, i.e., perturbation theory, is not effective because of the strength of the interaction.

All in all, theorists spent some 20 years in the dark. Finally, some new elements added to the renormalization theory paved the way for a solution.

In the next chapter, the strong interaction is discussed; the weak interaction will be treated later.

JULIAN SCHWINGER

He was born in 1918 in New York City. He was a prodigy and was discovered by Prof. I. Rabi of Columbia University; he wrote his first paper when he was 17 and took his doctorate at 21. In 1947, he developed quantum electrodynamics on his own and succeeded in explaining the Lamb shift, the anomalous magnetic moment of the electron, and so forth. As a result, he became a full professor at Harvard University at the age of 29, and he won the Nobel Prize along with Tomonaga and Feynman.

He, like Tomonaga, worked on the theory of wave guides during the war, has taught many students, and has had many collaborators. His significant contributions are many, but his work leans heavily toward mathematical physics. He is presently a professor at the University of California in Los Angeles.

16

QCD — THE QUANTUM MECHANICS OF COLOR

From Meson Theory to Color Dynamics

The purpose of Yukawa's meson theory was to explain the nuclear force. It is true that the nuclear force arises as the result of meson exchange; there are now, however, a countless number of different kinds of mesons, and it is impossible to derive the nuclear force quantitatively from a fundamental equation. But this is quite evident from the point of view of the quark model; describing the nuclear force is like describing the chemical properties of a complex modecule. Atoms are formed due to the Coulomb force between the nucleus and the electrons, but among neutral atoms the Coulomb interaction is replaced by more complex forces such as the van der Waals force and electron exchange force.

I have already introduced the idea that the hadron is a state of a quark atom that is neutral with regard to color. Color corresponds to the charge of the atom, and a Coulomb-like force acts among the colors of quarks. This kind of force might be simpler and more fundamental than Yukawa's nuclear force. But since there are three kinds of color as opposed to one kind of electric charge, the concept of charge must be expanded.

The theories of gauge fields, as they are nowadays called, represent a class of theories which share and generalize the characteristic properties of Maxwell's theory of electromagnetic fields. Such a generalization was first formulated by C.N. Yang and R.L. Mills, so the name "Yang-Mills field" is sometimes used for it. In a more extended sense, however, Einstein's theory of the gravitational field may also be called a gauge field theory. What, then, are the characteristics of a gauge field?

What is A Gauge Field?

First, it is, in general, true that the forces created by gauge fields are long-ranged and obey an inverse square relationship such as the Coulomb force and gravitation.

Second, the force is proportional to the quantum number of its source, and that quantum number is conserved. The electric charge of the electromagnetic field is an example.

These two characteristics are not only theoretically inter-related, but, more importantly, these requirements fix the equation of the gauge field; in this sense, it can be said that requiring the conservation of electric charge leads inevitably to the Maxwell theory. It is important, however, to distinguish between "Abelian" fields, such as the electromagnetic field, and other "non-Abelian" fields. The Yang-Mills field belongs to the latter category. Perhaps it is easiest to explain these ideas with gravity as an example; in the Abelian case, the field created by a source charge does not itself carry charge to become another source. But the fundamental requirement of Einstein's theory of gravity is energy equals mass, which means that all forms of energy become the sources of the gravitational field.

If a celestial object creates a gravitational field around itself, then the strength of the field is proportional to the mass of the object. But the gravitational field gives a potential energy to every point in space, so every point in turn must become the source of a gravitational field; this process must repeat itself indefinitely. The final result of the process is embodied in Einstein's equations which are nonlinear, as opposed to Maxwell's equations which are linear. This means that the gravitational field is not simply the sum of the fields due to each celestial body. However, its effect cannot easily be observed because, as a practical matter, the potential energy of the gravitational field is very much smaller in comparison with the masses of celestial objects.

The Dynamics of Color

Let us turn now to Yang-Mills-like fields. The theory of a Yang-Mills field with color as the quantum number is called chromo-dynamics; that is to say, the dynamics of color. By assumption, there are three colors — red, green, and blue — and the strong force acts between colored quarks; the hadrons are supposed to be a system in which the colors have cancelled themselves out and become white. How could such a theoretical system be realized?

It is not good enough to assign "electric charges" a, b, c to the three colors. What is required is white equals neutral, namely

$a + b + c = 0$,

but this would mean that the three charges are not equal. We can, in analogy to the process of expressing a mixture of primary colors, draw an equilateral triangle and assign each apex to a color; this, however, is not quantum mechanically satisfactory. The reason is that the symmetry regarding the mixing of the colored quarks as waves, that is to say the SU_3 symmetry, must be realized; this means that the process of, say, a red quark changing into a blue quark must also be allowed.

The result of all this is that eight kinds of gauge fields are needed. Each of these fields carries a mixture of color; the reason there are eight is the same as the reason for the Eightfold Way in the case of the SU_3 of flavor (u, d, s quarks). Let me explain it once again.

The Gluon — The Glue to Hold the Quarks Together

The quantum of the color gauge field is called gluon, meaning the glue that holds the quarks together.

Now let us think of a process in which a gluon is emitted by a quark. If, as a result of this process, the red (R) quark changes to a blue (B) one, then the gluon took red from the quark and gave blue. Equivalently, one can think of the gluon as having taken away red (R) and anti-blue ($\bar{B}$); thus this gluon is carrying a composite color of $R\bar{B}$. In general, the gluon, G_{ij}, that is released when q_i becomes q_j acts exactly like the compound state of q_i and $\bar{q}_j$ (Fig. 16.1).

$G_{ij} \sim q_i \bar{q}_j$

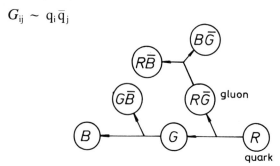

Fig. 16.1 Quarks have color and emit gluons.

There are $3 \times 3 = 9$ such combinations, and one of the nine gluons is a special combination corresponding to the color white:

$$G_W \sim q_R \bar{q}_R + q_G \bar{q}_G + q_B \bar{q}_B = 0.$$

But since it was required, to begin with, that the glue does not work on a white state, one must have $G_W = 0$. Thus the number of independent gluons must be eight.

Since gluons themselves carry color, the reader must be aware by now that there are possibilities of gluons themselves emitting gluons. Figure 16.1 expresses this phenomenon. It is the same circumstance as in the case of gravity and is what makes the gauge field of color non-Abelian. The word non-Abelian actually means "noncommutative". For example, if q_R emits gluons G_{RG} and G_{GB} in that order, it will first change to q_G and then to q_B; but the process in which the order of the gluons is reversed cannot occur; this is the meaning of noncommutativity.

Corresponding to the unit charge e of the electromagnetic field, the strength of the color carried by quarks is expressed by the unit g; this is a quantity not dependent on the kind of color. Just as the electromagnetic force from the exchange of a photon is proportional to e^2, the exchange of a gluon between two quarks should correspond to a force proportional to g^2. There are here, however, many possiblities of quarks changing or exchanging color depending on the kind of gluon involved. Furthermore, gluons can give rise to other gluons; there are also possibilities of exchanging more than one gluon. These give rise to forces that are proportional to g^4, g^6, . . . and so forth.

In the case of electromagnetism, $e^2/\hbar c$ is small so the Coulomb force proportional to e^2 took care of most of the forces, but in the case of gluons, the story is not so simple since we expect g^2 to be very much larger. Nevertheless it is still true that, in the cases where the state as a whole is white, the state is stabilized through attractive forces among the quarks and antiquarks

Asymptotic Freedom

QCD (quantum chromodynamics) refers to the quantum theory of color gauge fields. One can think of this theory as QED (quantum electrodynamics) with the electron replaced by the quarks and the photon by the gluons. But can this theory satisfactorily explain the strong interaction? In the first place, the process by which gluons are radiated, like light, from hadrons does not exist. It is not very convincing just to say that the colorless state has a tendency to be stable.

A theoretical progress occurred around 1973; it was the discovery of a strange characteristic called asymptotic freedom possessed by QCD. The discovery was made independently by t'Hooft (Holland), Gross and Wilczek (U.S.A.), and Politzer (U.S.A.); all four were either students or had received their degrees recently.

Asymptotic freedom can be thought of as one of the major characteristics of non-Abelian quantum fields (Yang-Mills fields) in general. In the case of QED, as was explained in Chapter 15, electron-positron pairs form clouds around a bare point electron which they tend to screen. Thus, the bare charge is larger than the measured charge. In the case of non-Abelian fields, the opposite situation prevails. Because of the fact that the fields themselves become sources of fields, like signed color "charges" form around the original charge, and the size of the charge becomes larger as one looks from farther away. Putting it another way, as one tries to find the true core charge by going through the cloud, the charge becomes smaller and smaller and fades away like a ghost. In terms of energy, it means that, as the energy is raised to probe the force at smaller distances, the force becomes gradually weaker than the Coulomb force because of the diminishing charge; this is the meaning of asymptotic freedom.

These strange things are the mathematical conclusions of the renormalization theory and are hard to understand intuitively. It is therefore not surprising that no one suspected these characteristics until the young people listed above actually made the calculations. (On the other hand, it is also true that there are considerable mathematical difficulties in applying renormalization theory to Yang-Mills field, and it took a long time to finally resolve them.)

Let us now return to QCD and discuss the significance of asymptotic freedom. It was originally thought that the field of force created by the color of the quark is Coulombic, but, as explained above, the strength of the color actually changes with the distance in such a way that the force is weak at short distances (high energies) and strong at long distances (low energies). It must be said that this characteristic is a very convenient one in explaining the behavior of quarks. That is to say, in high energy phenomena, quarks behave like free partons, and, at low energy, the interaction becomes very large.

Fig. 16.1 "Confined quarks".

If the interquark potential is as shown on Fig. 16.2 the result is the same as the string model and the quarks become confined. As we try to separate two quarks, there is no opposition at first; however, soon the glue becomes effective, and, unless the glue string breaks, the quarks cannot be forced apart. The same thing can be said about the gluons themselves. Since gluons also possess color, several gluons may come together and form colorless glueballs. Since these possess integer spin, they belong to the meson family; however they do not have any flavor and are thus presumably elusive. This may be the reason why there are no proofs of the existence of glueballs.

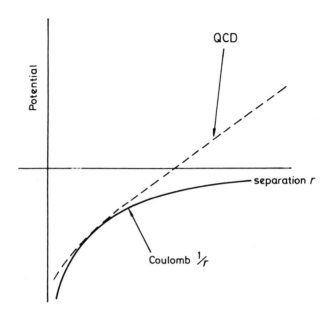

Fig. 16.2 The interquark potential compared with the Coulomb potential.

Such circumstances as outlined above led to the view that QCD is a very powerful candidate for the fundamental theory of quark dynamics. Still, not all the problems have been solved by any means. At present, the only area where precise predictions can be made are in the high energy phenomena where asymptotic freedom applies. Since the actual energies are not infinite, the quarks are not completely free; however, the corrections to the completely free state can be calculated. They are observed, for example, as deviations from the scaling laws and agree roughly with theory.

Let me here add an interesting fact. I said before that the asymptotic freedom arises because of the antiscreening characteristics of the gluons; however, the clouds that form around a single source are not only composed of gluons but also of quark-antiquark pairs. The latter produce the normal screening effect just as in the case of the electron-positron pairs in QED, so if there are more than 16 flavors of quarks, it turns out, that this screening effect overwhelms the gluon cloud and the asymptotic freedom is lost. We are safe, since there are only 5 or 6 known kinds of quarks; but is nature putting limits on the number of the kinds of quarks for such reasons?

Wilson's Lattice Theory

Let us turn now to the low energy phenomena. The first problem of QCD here is to find out if quark confinement occurs, and if it does, what the exact mechanism that governs this confinement is. But because the coupling constant is large, the mathematical problems become difficult, and a satisfactory method has not yet been found. However, in view of the fact that strings and bags have succeeded as models, there have been efforts to interpret these models in terms of QCD. Let me here describe Wilson's lattice theory as an example of such an approach.

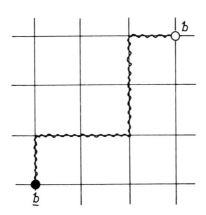

Fig. 16.3   An illustration of Wilson's lattice theory.

Lattice theory treats space-time not as a continuum but as a collection of discreet points, like crystals. According to Wilson's assumption, a quark occupies one of the lattice points and gauge field lines become the bridge between one of the lattice points to another. For example, in Fig. 16.4 quark q and anti-quark $\bar{q}$ are connected by a line of force (string), and the energy is proportional to the length of the string. There are infinite ways of choosing the path of the string, and a quantum mechanical average must be taken over these infinite ways. The result is that as the distance between q$\bar{q}$ increases, the energy becomes proportional to the distance, thus agreeing with the results of the string model.

17

SPONTANEOUS BREAKING OF SYMMETRY

What is Symmetry?

It is well known that the principles of symmetry play an important role in the laws of physics, and this book has periodically touched on this subject. If a symmetry (or an invariance) exists, then a corresponding conservation law follows. From the fact that space-time is invariant to parallel motion, the conservation laws of energy and momentum are created, and from the symmetry of direction, the conservation of angular momentum is born. Symmetries under rotation in space and under exchange of particle and antiparticle correspond to conservation of P and C respectively. The approximate symmetry of the u and d quarks leads to the approximate conservation of isospin, and so forth.

Thinking about what symmetry means in such cases as above we realize that it refers to a situation in which there exist several different states that, however, are basically equivalent to each other with respect to the laws of physics; symmetries imply that the laws of physics are invariant to the symmetry operations in which a state is replaced by an equivalent state.

But in any case, there is a need to distinguish between two equivalent states, say A and A'. To do this, it suffices to appropriately express the operation which changes A to A', and the quantities used in such expressions are the conserved quantities, or quantum numbers.

If physical laws possess a symmetry, existence of an arbitrary state A would imply that any equivalent state formed by a symmetry operation must also be possible. For example, if an apple is placed right side up on a table, the states where the apple is upside down, or somewhere else, are, in principle, also possible, and the characteristics of the apple are not thought to change in these other states.

There can exist, however, a state which is invariant to a symmetry operation, that is to say, a state that forms a group by itself. The mathematical sphere is invariant to rotations about its center, so talking about the direction of a sphere does not make any sense; a spatially displaced sphere, however, can be distinguished. On the other hand, the true vacuum, or the space-time in which there is no matter, is thought to be invariant under any symmetry operation. (Homogeneity and isotropy of space-time: p. 81) Thus, there is only one vacuum; it is the placing of the matter in the vacuum that causes the diversity, and at the same time the increased energy, of the world.

When we attempt to characterize the state of the entire Universe with the quantum theory of fields, we assume that the vacuum is a unique state which is invariant to all symmetry operations, based on the above considerations. Of course, the same postulates should apply to a large enough vacuum, though not the Universe itself, for which the influence from outside the vacuum area can be ignored.

Spontaneously Broken Symmetry

The above asumption about the vacuum, however, is only an assumption and is not a self-evident fact. Let us illustrate this with the rather well-known example of ferromagnetic medium. Ferromagnetic materials are those materials, such as iron or nickel, that can become permanent magnets. The reason these particular materials can become magnetic is that each of the atoms, which are small magnets due to the spin of the electron, interacts with adjacent atoms in such a way that the energy is lower when they are lined up. Thus, a macroscopically magnetic state where all of the small atomic magnets are lined up is the most stable ground state. But exactly in which direction the atoms choose to line up is arbitrary; at least the problem can be ideally thought of in this way. Thus, this medium is invariant to the rotation of the direction of all the atoms, and the number of ground states, corresponding to the number of possible directions, is infinite.

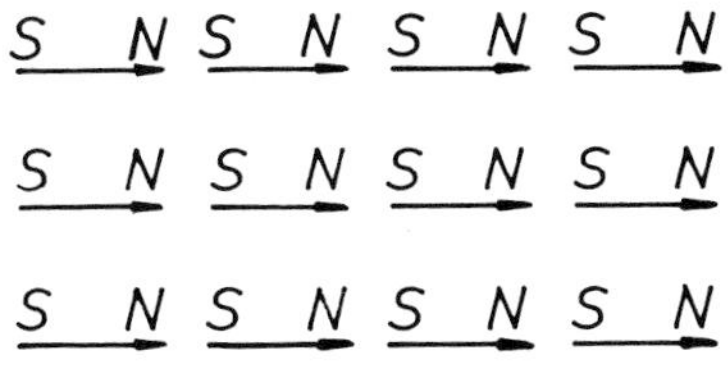

$$S \quad N \quad S \quad N \quad S \quad N \quad S \quad N$$

$$S \quad N \quad S \quad N \quad S \quad N \quad S \quad N$$

$$S \quad N \quad S \quad N \quad S \quad N \quad S \quad N$$

Fig. 17.1 Ferromagnetism occurs when all atomic magnets face the same direction.

If the size of the magnetic material is finite, there is no problem with this infinity. If there is a magnet with a particular direction, we can make, in principle, the same magnet with a different direction, and these two can also exist together. But what would happen if the magnet had an infinite extent. We would always live in the magnet, and the ground state must be chosen in some direction.

In the case of an infinite magnet, even though there is no energy required in changing the direction of the ground state, there is no way to carry out such an operation, the reason being that such a change of direction would require the change of direction of all the atoms that compose the infinite medium, which is impossible with a finite device. The only thing we could do would be to change the direction of a portion of the medium; but if we did this, the medium would develop a local disturbance and would no longer be in the ground state. Thus we would live in a world with a fixed direction where only the local disturbances of order are observable. For observers in such a world, symmetry becomes no longer self-evident.

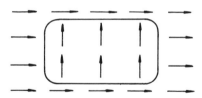

Fig. 17.2 Spontaneous symmetry breaking: Only local disturbances of order are observed.

The above phenomenon is called spontaneous breaking of symmetry. As can be seen by the present example, effectively infinite degrees of freedom are needed for spontaneous symmetry breaking to occur; however, there are many example of this mechanism in everyday physical phenomena. The first time that this mechanism came to be perceived as a general principle, however, was when the Nambu-Jona-Lasinio theory (super-conductivity model) applied this principle to particle theory.

Let me here present an analogy of the spontaneous symmetry breaking, first used by A. Salam. Let us say that there is a banquet and many people are sitting around a large round table. Each seat has a plate, knife, fork, napkin, etc., but since the settings are so close together we can't tell whether the napkin on the right-hand side or the left-hand side belongs to a seat; in other words, there is a right-left symmetry. It really does not matter which of the napkins is taken; however, when someone takes the napkin on the right-hand side, everybody else must also take the napkin on the right-hand side; the symmetry is suddenly spontaneously broken.

Spontaneous symmetry breaking — Salam's analogy.

The Relics of Spontaneous Symmetry Breaking

The instance in which symmetry is spontaneously lost and the instance in which there was no symmetry to begin with are not necessarily equivalent. In the case of the permanent magnet, the symmetry concerns the continuous and infinite number of directions of the magnets; in the case of the banquet, the symmetry concerns the finite, discontinuous possibilities of left and right. In such cases as the former, with continuous symmetries, the spontaneous breaking of symmetry leaves some relics.

As stated above, it is difficult to perform the operation which twists all of the atoms in the magnet at the same time. If, however, such a twist is performed in a finite region, a twist wave (spin wave) with a wavelength the size of that region should occur. But if the size of the region is increased to infinity, the operation approaches that of the original symmetry operation under which the corresponding energy of the wave should be zero. The characteristics of such waves are the same as those of light or sound; which is to say that the quantum of the wave does not have a mass. In other words, quanta of arbitrarily small energy exist, which are the relics of the broken symmetry.

In general, such waves are called Nambu-Goldstone (NG) waves. In fact, the elastic wave that occurs in solids is an example; this can be thought of as the result of the breaking of the symmetry of parallel translations in the solid. An atom, in principle, can occupy any point in space; in fact, in a crystal, the choice of the position for the first atom forces all the other atoms to sit in an order relative to this position. The attempt to disturb this order creates the elastic waves.

Superconductivity

Superconductivity is the phenomenon where the electrical resistance of a material becomes completely zero under a certain temperature, and the material also expels any magnetic fields; there are many substances that exhibit this behavior at low temperatures. Bardeen, Cooper, and Schreiffer (University of Illinois) succeeded in theoretically explaining the mechanism of superconductivity, and their theory is popularly called BCS theory.

Superconductivity is not only interesting in that the symmetry broken here is an abstract one of gauge invariance (conservation of charge), but it is also very important as a model for particle theory.

Cooper pair liquids flow smoothly.

It is not the purpose of this book to present the BCS theory; but let me quickly summarize the theory here. In general, electrical conduction occurs when some of the electrons in a material are able to move about freely; in the case of superconductivity, however, pairs of spin up and spin down electrons (called Cooper pairs) condense out in the material. Superconductivity can be interpreted by saying that it occurs because the number of pairs is not rigidly determined, and thus the state does not change even if the pairs float around from place to place.

On the other hand, it takes an energy greater than some minimum energy E_0 to break a Cooper pair, and the single electrons that result from this breaking behave as though they were particles of rest mass $E_0/2$. This mass is different from the normal mass of the electron, and the extra mass is added on only inside the superconductor.

The above situation is very rich in possibilities of application. What would happen if a kind of superconducting material occupied all of the Universe, and we were living in it? Since we cannot observe the true vacuum, the ground state of this medium would become the vacuum in fact. Then even particles which were massless (for example, the neutrino) in the true vacuum would acquire mass in the real world. Perhaps the masses of electrons and quarks were acquired in this way.

The Nambu-Jona-Lasinio (NJL) superconductivity model was the first to pursue the above analogy; but before this model can be explained, the relationship between superconductivity and spontaneous breaking of symmetry must be clarified.

The symmetry broken in the BCS theory concerns the conservation of charge. The symmetry is broken because the number of Cooper pairs are not fixed so the total charge is somewhat smeared; but the smear is not random but obeys an order. This order is specified by a parameter, which corresponds to the direction of spin in ferromagnets, and has a certain amount of freedom.

The Origin of the Quark Mass

Let us now return to the problem of elementary particles.

Since the NJL model came before the quark model, it considered the dynamical origin of the mass of the baryon assuming that the baryon corresponded to the electron in a superconducting body; however, the model would apply equally well to the masses of today's leptons and quarks. In any case, the symmetry in question does not concern electric charge but something called chiral symmetry.

Chiral is a term that refers to the distinction between right-handedness and left-handedness. It is known experimentally that the neutrino spin can come in right-handed and left-handed varieties (with respect to the direction of motion). These two are to be referred to as having chirality ± 1.

In the real world, only the left-handed neutrino ν and the right-handed antineutrino ν exist, which means that parity is broken, as explained before. But such situation is peculiar to massless neutrinos; if a particle has mass, both right-and left-handed components must exist. The reason is that massive particles must travel slower than the speed of light. Even if a particular particle is left-handed with respect to its direction of motion, the particle looks right-handed from a reference frame that goes past the particle, since the direction of motion is reversed.

Fig. 17.3 Left-handed spin looks right-handed when one looks while going past the particle.

The fundamental characteristics of the weak interaction is expressed by something called V minus A theory. According to this theory, the weak interaction, in general, involves only the left-handed component of the particles and the right-handed component of the antiparticles. Also, since the particles do not change into antiparticles, the total chirality, that is to say the number of left-handed particles minus the number of right-handed particles, does not change as the result of the interaction. But since particles other than the neutrino have mass, the left and right are mixed when these interact, so the conservation of chirality is broken anyway.

The above should make the analogy with superconductivity clear. We say that leptons and quarks do not have mass to begin with but, in the real world, the symmetry is spontaneously broken, and they appear as particles with mass. But this is only a possibility; we need some other theoretical support. This is provided, fortunately, by the NG wave.

Let us now explain the correspondence to the BCS theory in more detail. In this world, particles and antiparticles (for example, q and $\bar{q}$) condense out as Cooper pairs of chirality zero. Breaking one of these pairs produces a massive quark and a massive antiquark. Disturbing the distribution of the pairs creates NG waves; these waves have spin zero and parity minus, the same quantum numbers as those of, say, π-mesons. Perhaps the reason the pion has a much smaller mass than other hadrons is because it is precisely the quantum of the NG wave.

In reality, there are three kinds of pions, and also the much heavier K-mesons and eta (η) mesons have the same spin and parity as the pions. To treat all these mesons in the same way, quark flavors must be included in the chiral symmetry, and also we must assume that the masses of these mesons are not zero because they are not entirely due to the mechanism of spontaneous breaking of symmetry. The last point is rather unsatisfactory and has not been completely resolved; however, the spontaneous breaking of chiral symmetry has been successful not only as an interpretation of hadron physics; the concept of spontaneous symmetry breaking has become an important basic concept in the unification of gauge fields that will be discussed later.

18

THE LEANING
STRUCTURE OF THE
WEAK INTERACTION

God's Mistake?

It is rather difficult to find out the true identity of the weak interaction. This is not merely because the coupling constants are small and the range is short for the weak interaction. Whereas the other interactions — gravity, electromagnetism, strong force — all fit in the framework of gauge theories and obey beautiful symmetry principles, the weak interaction seems to be very unstructured and has no perfect symmetries. I already mentioned the breaking of parity (P), CP, and strangeness in Chapters 6-8; the prevailing feeling is that the breakings of symmetries have not all been found yet. Let me here state what I always feel when I think about the problem of the weak interaction.

Is the weak interaction God's mistake?

When God made the plans of the Universe, He accurately drew in the structure of gravity, electromagnetism, and strong force. But when He came to the weak force He found that, perhaps because of mistakes in reading a ruler or because of miscalculations, there were discrepancies in the drawings. Straight lines did not cross at right angles, and rectangles did not close right; and the whole structural frame of the weak interaction was at an angle with those of other forces. But since the discrepancies were not so noticeable from a distance, God went ahead and built the Universe.

Scientists, however, believe that the omnipotent God does not do such sloppy things, and they try to find the explanation for all phenomena based on this belief. We assume that the reason the structure is leaning is not the result of some mistake but is based on some necessary reason.

Persistent and strenuous efforts at finding this necessary reason have been made in the last 20 or so years, and, as a result, our understanding of the weak interaction has increased enormously. The weak interaction has been systematized by the Weinberg-Salam theory and can now be written down in a unified manner. It has become possible to interpret the weak interaction on the basis of gauge fields like the other interactions. But before this theoretical development can be followed, we must explain what the leaning structure of the weak interaction is.

What is the Weak Interaction?

The weak interaction originally meant the beta decay of the nucleus, but, as we learned earlier, its domain has been expanded to new particles, and there is a rich variety of phenomena under the weak interaction. For example:

$$n \rightarrow p + \bar{\nu}_e + e, \tag{a}$$

$$\Lambda \rightarrow N + \pi, \qquad (N = p, n), \tag{b}$$

$$\Lambda \rightarrow p + \bar{\nu}_e + e, \tag{c}$$

$$K \rightarrow \pi + \pi, \tag{d}$$

$$K \rightarrow \pi + \pi + \pi, \tag{e}$$

$$K^- \rightarrow \mu + \bar{\nu}_\mu, \tag{f}$$

$$\pi^- \rightarrow \mu + \bar{\nu}_\mu, \tag{g}$$

$$\mu \rightarrow \nu_\mu + e + \bar{\nu}_e, \tag{h}$$

$$\mu + n \rightarrow p + \nu_\mu, \tag{i}$$

and so forth. There are various kinds here such as those that involve only leptons, those that involve only hadrons, and those that involve both leptons and hadrons, and it would seem very difficult to try to grasp all these in a unified way. The matter is simplified considerably, however, by going down to the level to the fundamental particles, i.e., the leptons and quarks. For example, the reaction (g) which is a π^- decay can be thought of as:

$$\pi^- = (d\bar{u}), \qquad d + \bar{u} \to \mu + \bar{\nu}_\mu.$$

The decays of neutrons and muons (a, h, i) can be rewritten as:

$$n + \bar{p} \to e + \bar{\nu}_e,$$
$$\mu + \bar{\nu}_\mu \to e + \bar{\nu}_e,$$
$$n + \bar{p} \to \mu + \bar{\nu}_\mu,$$

which are of the same form. Indeed since p and n are composite particles udd and uud, all of the above three reactions can be written as the quark reaction:

$$d + \bar{u} \to e + \bar{\nu}_e,$$
$$\to \mu + \bar{\nu}_\mu.$$

Thus, the weak interaction becomes the process of exchanges among the quark or lepton pairs. (Reactions such as $K \to 3\pi$ are more complex, but they can be understood as involving the strong interaction as well.) It is natural to expect that there is a regularity to these exchanges; but is there really a regularity?

Orderliness in the Weak Interaction

Historically, the theory of weak interaction begins with the Fermi theory of 1933. In Fermi's theory of beta decay, an electron-neutrino pair is created, when the neutron decays into a proton; (p. 185 Eq. (a) but this is thought to be a point interaction, that is to say, a reaction that occurs at a point in space.

In general, the features that characterize an interaction are its strength (coupling constant), and type (vector type, scalar type, etc.). The strength determines the rate of the decay reaction (or the lifetime), and the type determines such things as the energy distribution of the electron, whether the spin of the neutron changes direction during the decay, and other subtle effects. Fermi hypothesized the weak interaction to be of the vector type, which does not change the direction of spin; however, this contradicted the results of some experiments, so G. Gamow and E. Teller added another type of force. It was in 1957 that the non conservation of parity was discovered and the so called V minus A type was established as the correct type of interaction. The interaction is thought to be an almost equal mixture of vector (V) and pseudovector (A) types, and the coupling constant is called the Fermi constant.

It is easy to apply the V minus A theory to all quarks and leptons. As was shown above, all fundamental particles must be paired like (e, v_e), (μ, v_μ), (u, d) and so forth, and any of these pairs now needs to be combined into a pair of pairs. The muon decay occurs through the combination (μ, v_μ) $\times$ (e, v_e), the neutron decay (a) through (u, d) $\times$ (e, v_e), the π-meson decay (g) and the μ capture (i) through (u, d) $\times$ (μ, v_μ). If all of these decays have the same coupling constant (Fermi constant), there should be a set relationship in their lifetimes. Except in the case of the π decay, for which the calculations are difficult, the other three, indeed, have the regular relationship that is expected.

It has thus become clear that the weak interaction also has order, but there still remains one difficulty; this concerns the strange particles. For example, the lambda particle (Λ) decay (c) closely resembles (a), but it must be thought to involve the pair (u, s). However, the d quark is already paired with the u quark in (u, d). Furthermore, the coupling constant for (u, s) is only about a fourth of the Fermi constant. The situation is the same with the other strange particles Σ (sigma), Ξ (xi), and K.

There must be a Charmed Quark

To resolve the problems outlined above, N. Cabibbo (University of Rome) proposed the following ingenious idea. Instead of taking two different pairing (u, d) and (u, s), the pair (u, d ') is taken; d ' is a mixture of d and s. (This is called Cabibbo mixing. Thus, we mix the quarks (i.e., their waves) of different flavor but of the same charge.) Then, the probabilities of the transitions within each of the pairs (u, d) and (u, s) become smaller than normal, but if d ' is mostly d, the (u, d) transition probability will be roughly unchanged, whereas the (u, s) transition probability will be much smaller.

According to the above theory, all strange particle decays should be understood, once the mixture of d quark and s quark is fixed. Experiments confirmed these expectations, and the Cabibbo theory was a success. But why does a funny mixture like d ' appear in the weak interaction? Perhaps the problem is in trying to force pairs when there is an SU_3 symmetry among the three flavors (u, d, s). Such questions were raised by people like Z. Maki and Y. Hara, who argued as follows.

From the standpoint of weak interactions, it would be natural to expect the existence of a pair (c, s ') along with the pair (u, d ') . If, here, the d ' is a mixture of a d quark and an s quark, then the s ' would be a "perpendicular" mixture which would consist mainly of the s quark. The c would be a new fourth quark and would carry electric charge $\frac{2}{3}$ just like the u quark, but we would assume that it has not been discovered yet (as of that time). because its mass is very much larger than the u quark. Of course, this is exactly what we call the charmed quark today.

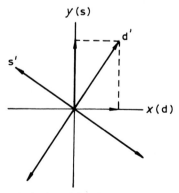

Fig. 18.1   The polarization of light is analogous to Cabibbo's theory.

To understand the relationship between the pairs (d, s) and (d ', s '), it is best to draw an analogy with the polarization of light. As can be seen on Fig. 18.1, a linearly polarized wave can always be decomposed into the perpendicular components of x and y. If these two components symbolized the d quark and s quark, d ' would correspond to one polarization in the figure, and s ' would correspond to the polarization perpendicular to d '. It is also possible to decompose the d and s quarks into the direction of d ' and s '; only the choice of the coordinate axes would be different.

The above situation is what I meant when I said that the structural frame of weak interaction is leaning. In the strong interaction, different types (flavors) of quarks are distinguished along the d and s axes, whereas the weak interaction distinguishes them in the d ' and s ' axes.

Maki's theory was not the only one that predicted the existence of the charmed quark; however, it is worth pointing out that his theory was firmly based on a guiding principle of Sakata's — that there need not be any limits on the number of elementary particles. Later, it was shown by S.L. Glashow, J. Iliopoulos, and L. Maiani (GIM) that the charmed quark is also necessary to solve the problem of the so-called neutral currents, so the necessity of the existence of charmed quark was even more strongly felt.

I will explain the neutral current in some detail later; but in the usual beta decay, u $\longleftrightarrow$ d ', e $\longleftrightarrow$ ν, a charge is always exchanged between the pairs. In neutral current interactions, on the other hand, the charges are not exchanged. An example would be

$$d + e \rightarrow d + e.$$

The existence of such reactions is demanded from the gauge theory of weak interactions, but the above reaction cannot be directly observed as a decay process of the d quark. On the other hand, the process s $\longleftrightarrow$ d is also contained in the transition d ' $\longleftrightarrow$ d ' in the Cabibbo frame; this would cause the decay

$$\Lambda \rightarrow n + e + e$$

which should occur with the same frequency as the charge exchange reaction $\Lambda \rightarrow p + e + \nu$. But the former reaction is not actually observed. The GIM theory explains this by saying that s $\longleftrightarrow$ d is included in s ' $\longleftrightarrow$ s ' as well as d ' $\longleftrightarrow$ d ' and they tend to cancel each other out.

The leaning structure concept started by Cabibbo became further developed by M. Kobayashi and T. Maskawa. Up to now, we considered only the pairs of quarks (u, d) and (c, s), and the leaning appeared as the mixing of d quark and s quark. However, the same result is obtained if the u quark and c quark are mixed instead, so the choice is arbitrary. But when the number of quark pairs increases to three, the situation becomes different.

Let us name the third pair (t, b). This time, the d quark, s quark, and b quark can mix together and become d ', s ', and b '. It was pointed out by Kobayashi and Maskawa that there is a larger freedom in the mixing of the three pairs than in that of the two pairs so that there is a possibility that CP may not be conserved. In other words, they suggested that the fact that CP is broken in the real world may be an indication that there are at least three pairs of quarks.

It is rather amazing that all of the above thinking predates the discovery of the charmed quark. The Kobayashi-Maskawa theory can be said to have been substantiated by the recent discovery of the upsilon particle (Υ).

Why the structure of the weak interaction is tilted with respect to the other interactions is an unsolved mystery. It is rather interesting, however, to imagine what would happen if there was no tilt, and God's design was perfect. Leptons and quarks would come in pairs such as (e, ν_e), (u, d), (μ, ν_μ), (c, s) and so forth, and changes would occur within a pair but never among different pairs.

These pairs of leptons and quarks look the same except for their masses. Actually we do not know how many pairs there are in nature, so they are sometimes called generations. (e, ν), (u, d) is the first generation, (μ, ν_μ), (c, s) the second generation, and so forth. If the generations do not mix, generation number would become a kind of conserved quantity. For example, the change c $\rightarrow$ s may occur in the second generation but the s quark could not change into the first generation u quark. So particles like the lambda (Λ) and K would be stable; the nuclei, in general, would be composed of proton, neutron, lambda particle, K-meson, etc., and the world would be a very different place.

19

THE WEINBERG-SALAM
THEORY

Beyond the Yukawa Meson Theory

If the electromagnetic interaction can be described by quantum electrodynamics (QED) which begins with the Maxwell theory, and the strong interaction by QCD based on the gauge field of color, what is the gauge theory that describes the weak interaction? The answer to this question is given by the Weinberg-Salam (WS) theory.

How well Fermi's theory could be expanded to include all of the fundamental particles was explained in the last chapter; however, this point interaction theory (p. 185) is unnatural from the point of view of the Yukawa concept of the interaction that is mediated by a field. In fact, when Yukawa developed the meson theory, he also tried to explain beta decay with the same theory. Thus beta decay would involve the steps:

I. $n \rightarrow p + Y^-$,

II. $Y^- \rightarrow e + \bar{\nu}$.

Beta decay was thought to occur as an exchange of the meson Y between the nucleon and the lepton just as the nuclear forces arise from the exchange of the Y-meson among the nucleons. But this thinking quickly encountered a problem. This scenario contradicted the fact that the coupling constant for the neutron decay and the muon decay was the same (p. 187).

In the Yukawa two-step hypothesis, the first step (I) in the neutron decay should be the same as in the case of the nuclear force so the coupling constant here should be large; this would mean that the second step (II) is weak. But in case of the μ, which is a lepton, there should be no strong force involved in the decay so both steps must be weak. Thus the universality of the coupling constant cannot be explained. All of this leads to the conclusion that if there is a field that mediates the weak interaction, it is different from the meson, and it is something that couples to both leptons and hadrons (or quarks) without distinction.

The W Boson

The quantum of this new field is called the W boson. (W stands for weak.) The theory may be thought of as applying the Yukawa theory of nuclear force to the weak interaction.

Let us once again write down the beta decay of the neutron and the muon:

S. L. GLASHOW   **A. SALAM**   **S. WEINBERG**

These three men jointly received the Nobel Prize in 1979 for their contribution to the unified theory of weak and electromagnetic interactions. Glashow and Weinberg are presently professors at Harvard University and the University of Texas, respectively.

Glashow is also well known for the GIM (Glashow-Iliopoulos-Maiani) theory, which predicted the existence of the charmed quark, and the Georgi-Glashow SU_5 theory that was the beginning of grand unification theories.

The areas of contributions made by Weinberg are wide and include particle physics and cosmology; his efforts in bringing these two fields together have been important. He is also known for his writings, like *The First Three Minutes*.

Salam was born in Pakistan. While serving as a professor at London's Imperial College, he has spent much of his efforts in supporting physicists of developing countries at the International Center for Theoretical Physics at Trieste which he built. He is also very energetic in his research activities; he was recently been making contributions in supersymmetry and other frontiers of field theories.

$$n \rightarrow p + W^-, \quad (\text{or } d \rightarrow u + W^-)$$
$$\mu \rightarrow \nu_\mu + W^-,$$
$$W^- \rightarrow e + \bar{\nu}_e .$$

The above equations and those with charges conjugated will indicate that the W (and its antiparticle) is a boson with the electric charge ± 1.

What are the mass and the spin of the W boson? The question of spin is simple; for the beta decay to be a V minus A type interaction, the W boson must have spin 1 and must couple to the left-handed components of the quarks and leptons (and the right-handed components of their antiparticles).

In other words, the V–A universal interaction is created because the spin of the W boson is 1 and because these particles couple with the same strength to all particles. This corresponds to the fact that the spin of the photon is one and that the electric charge is universal, and suggests that the weak force is also due to the gauge field. However, the W boson lacks the characteristics of the gauge field in other respects.

As opposed to the electromagnetic force which is a long range force whose quantum (photon) has mass zero, the range of the weak interaction is extremely short and was assumed to be zero in the Fermi theory. This last assumption is equivalent to assuming the W boson mass to be infinite. If the range is finite, however, the Fermi constant is expressed as the square of the product of the coupling constant g of the W boson and the range of the interaction (or equivalently, the square of the product of g and the inverse of the mass M). Thus it is impossible to determine both g and M just from the Fermi constant alone.

The relationship above only applies to low energy phenomena when the de Broglie wavelength of the particle undergoing the interaction is much larger than the range of the force; the beta decay is such a phenomenon. On the other hand, it is also possible to have the weak interaction at high energies.

A good example of a high energy weak interaction is the neutrino reaction.

$$\nu_\mu + n \rightarrow \mu + p + \ldots$$

Fig. 19.1 Neutrino facility at CERN (Geneva). (*Courtesy of CERN*)

Here, high energy neutrinos hit a target and the outgoing particles, such as muons, are detected. This reaction is similar to the electron-proton electromagnetic scattering; the W boson is exchanged rather than the photon.

According to Fermi theory, the cross section of the above reaction should increase in proportion to the energy; but if the mass of the W boson is finite (if the range of the weak interaction is not zero) then the cross section should stop increasing at some point. Up until now, there has been no evidence for the falling-off of the increase of the cross section, so, unfortunately, it must be interpreted that even if the W boson exists, its mass is too large for us to measure.

Incidentally, the neutrino experiment discussed above is by no means easy to do. Even though it is convenient for large accelerators that the cross section increases with the energy, the cross section is still quite small in comparison with those for the strong interaction.

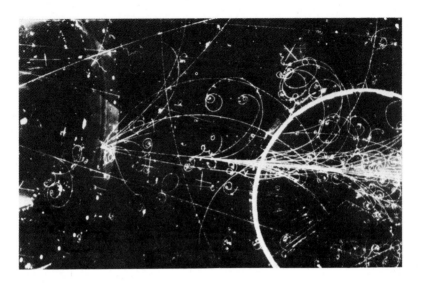

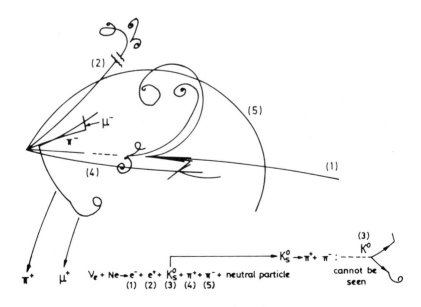

Fig. 19.2   The target (Ne) is struck by a neutrino and the outgoing particles are examined. In this reaction the ν_e collision with Ne creates a charmed hadron which decays into K^0.

To make a neutrino beam, first, a proton beam must strike a target to make π-mesons. The neutrinos from the decays of the pions and K-mesons, that are produced in the forward direction, interact in the detectors such as bubble chambers. However, it is difficult to get rid of unwanted particles from the beam; and the pions do not decay easily because their lifetime lengthens with increased energy. So the neutrino beam at Fermilab runs for hundreds of meters through earth before it strikes the detector.

The Unification of Weak and Electromagnetic Forces

Even if the existence of the W boson has not been experimentally proven, can its mass be predicted theoretically? As pointed out earlier, this would be possible if the coupling constant for the weak interaction is known. What would happen if we assume that the coupling constant is the same as that of the electromagnetic field (i.e. unit charge)?

It is natural that such thoughts as above should come to peoples' minds. The theory of the unification of the electromagnetic force and the weak force indeed starts from such a simple motivation, but, of course, many other arguments can be pasted on to the basic idea. For example, the weak and electromagnetic forces act on both quarks and leptons equally. Thus, unlike the strong force that only couples to the colors of quarks, weak force acts on the flavor quantum numbers of the quarks and leptons. Perhaps the reason that the weak interaction is very much weaker than the electromagnetic interaction, as well as having completely different characteristics, may be simply due to the mass difference between the photon and the W boson.

If the coupling constant were the same as the unit electric charge, the mass of the W boson would be 40 GeV. In comparison with the masses of hadrons which are about 1 GeV, the W boson must be called a very heavy particle; it is understandable that it has not been found yet. On the other hand, since weak interaction breaks parity, for example, perhaps it cannot simply be described by the same coupling constant as the electromagnetic force.

There had been attempts at electroweak unification theories along the above outlines by Glashow, Salam, and others from early on, but these theories cannot be called real gauge theories. It was in 1967 that a logically complete theory was introduced by Weinberg and Salam, independently, and this is what is today called the Weinberg-Salam (WS) theory.

Is mass the difference between the photon and the W boson?

Comparison with Superconductivity

The WS theory, in many respects, takes superconductivity, previously given as an example of spontaneous breaking of symmetry, as its model. Let us attempt an explanation of the theory in terms of similarities to the phenomenon of superconductivity.

Insulators are "transparent" to electric and magnetic fields and they allow the fields to penetrate. Normal conductors are transparent to magnetic fields but expel electric fields. Superconductors, on the other hand, expel both electric and magnetic fields. (The expulsion of magnetic fields is called the Meissner effect.) This expulsion, however, is not complete and the fields can reside in the thin skin at the surface.

The difference between the characteristics of transparency and expulsion is the same as that between the Coulomb force and a Yukawa type force, and it indicates that the range of the field can become either finite or infinite depending on the medium. The skin depth to which the field penetrates is the range in that medium. So if the Yukawa theory is applied to the present situation, there should exist a "particle" whose mass is inversely proportional to the skin depth.

Thus, in a transparent medium, the photon is a massless particle, but in a superconductor, it will act as a particle with a fixed mass (rest energy); such a particle is called a plasmon. If we lived in a superconductor, the electromagnetic force arising from the exchange of plasmons would appear as a Yukawa type force.

Since the WS theory attempts to unify electromagnetism and weak interactions, the real situation in much more complex. Besides the $W^{\pm}$ bosons and the photon, there is a boson called Z^0 in the theory which is responsible for the weak neutral interaction; that is to say, interactions such as $\nu \to \nu$ or $e \to e$ in which there is no change of electric charge. For example:

$$\nu + p \to \nu + p, \qquad \bar{\nu} + p \to \bar{\nu} + p.$$

Whether such neutral interactions exist at all is an interesting question in itself, so a series of neutrino beam experiments were carried out in both Europe and America. These experiments established the existence of these interactions.

Let us now return to the WS theory. This theory is sometimes called the $SU_2 \times U_1$ gauge theory because it is made up as a combination of the non-Abelian SU_2 and the Abelian U_1 gauge fields, which, in general, have different coupling constants. The meaning of the above can be seen by examining the way these gauge fields couple to quarks and leptons.

Pairs such as (u, d) and (ν, e) are seen as possessing "weak" isospin $\frac{1}{2}$, and the transformations among their components are the "weak" SU_2. On the other hand, the U_1 has to do with the quantum number that distinguishes the quarks and leptons. The SU_2 gauge field has the "weak" isospin 1 with three components, and the U_1 gauge field has "weak" isospin 0. In comparison with the strong isospin, these gauge fields are something like the π-mesons ($\pi^{\pm}$, π^0) and the η^0-meson. The $\pi^{\pm}$ corresponds to the $W^{\pm}$ bosons and the appropriate mixture of the π^0-meson and the η^0-meson correspond to the Z^0 and Y. (Here, "mixture" has the same meaning as the mixture of the s quark and d quark of Cabibbo (p. 188)).

To divide the four gauge fields into an electromagnetic field γ and the weak fields $W^{\pm}$, Z^0 and to give mass to the latter requires the world to be a rather complex "superconductor" I will not go into the actual mathematical tricks, but what represents the "superconductor" in the WS theory is something called the Higgs field. (The method used here is the same as that used by the famous Russian theoretical physicist, L.D. Landau, for real superconductors.) The Higgs field and the weak field, like oil and water, do not mix; this corresponds to the Meissner effect.

The WS theory was introduced in 1967. The reason the theory was not immediately accepted was not only because the existence of the neutral interaction had not yet been proven but because there were questions as to whether this theory fell within the framework of the renormalization theory. If renormalization is impossible, the theory is not a useful gauge theory. It was only after t'Hooft of Holland proved, in 1973, that renormalization was possible that the WS theory has come to be seen as equal to electrodynamics.

WS theory has been so far successful in describing the known phenomena; the real verification, however, will require the experimental confirmation of the existence of the quanta of the gauge field $W^{\pm}$, Z^0 and the quantum of the Higgs field H.

The masses of the $W^{\pm}$ and Z^0, though not identical, are both calculated to be in the neighborhood of 90 GeV. The reason this value is different from the naive estimate of p. 199 is because the WS theory has two coupling constants, and their relationship to the electric charge is somewhat complex. In any case, it probably is not in the far future that these particles will be created in the laboratory.

The confirmation of the $W^{\pm}$ and Z^0 particle came in 1983 when these particles were created in the proton-antiproton colliding beam experiments at CERN. C. Rubbia and S. van der Meer, the two people who were instrumental in carrying out the project, have just received the Nobel Prize.

20

THE UNIFIED FIELD THEORY

Unification of the Three Forces

The fact that the Weinberg-Salam theory found order in the weak interaction and gave it, as a renormalizable gauge field theory, a position equal to that of electrodynamics was a remarkable theoretical progress. Furthermore, the theory treats electromagnetic and weak fields not as separate entities but as things which are intimately connected within the same framework.

Although the coupling constants of the electromagnetic force and that of the weak field W are not the same, they are very close.

We also have a gauge field theory that is applicable to the strong interaction — QCD (quantum chromodynamics) — the quantum mechanics of color. But the coupling constant of the color field (gluon) is very much larger than that of the electromagnetic field. Furthermore, it was explained earlier (p. 168) that the coupling constant of the strong field cannot be expressed as a fixed value but must change due to the energy, or length, scale at which the measurement is done.

That the coupling constant changes according to the scale of observation is not a peculiarity of QCD but is a general characteristic of renormalizable theories. In the case of Abelian gauge theories such as QED (quantum electrodynamics) the behavior of the coupling constant is opposite that of the non-Abelian fields and it rises with the energy. It was explained already that the electric charge looks larger at closer distances and looks, because of screening, smaller at longer distances. The screening action is not complete, however, so the charge seen at an infinite distance is not zero but possesses a fixed value; this value is what we normally mean when we talk about the size of the charge.

Now, let us think about the strong, electromagnetic and weak forces all at the same time; can all of the forces be unified under one theory?

The observation made above seems to strongly support the possibility of unification. This is because, in the case of the coupling constant of the electromagnetic field, the initially small coupling constant rises with energy; the coupling constant for the strong force, on the other hand, is large but becomes smaller with increasing energy; the weak interaction coupling constant also should become smaller since the field is non-Abelian.

Thus, perhaps the coupling constants of the three kinds of interactions come together at some energy, or length scale. Then perhaps we can say that there is really only one coupling constant, and the three gauge fields are simply components of a single gauge field. Perhaps the phenomenon of spontaneous symmetry breaking forces these components to evolve along different paths and appear in different forms at low energies.

Incredible Energy — But We Can't Say It's Nonsense

If the above idea is true, it is a wonderful thing. But what is the energy at which this unification occurs? To examine this question, the following characteristics of the renormalization theory must be presented.

Normally, the coupling constant g of a gauge field is expressed by the dimensionless number g^2/hc. In the case of the electromagnetic field, this number is the fine structure constant $e^2/hc = \frac{1}{137}$. In QCD, the gluon coupling constant g^2/hc, at around the energy of 1 GeV, is known to be about 0.3. These coupling constants, in general, can be thought to be inversely proportional to the logarithm of the energy E.

The constant of proportionality is fixed by the details of the theory; in the case of non-Abelian fields it is negative, in the case of Abelian fields it is positive. The rate of change of the coupling constant is quite slow since the change is logarithmic. The value of energy at which the coupling constants of the electromagnetic field and the strong field come together, computed by using the above relationships, turns out to be an incredible number of 10^{15} GeV. This is 10^{13} times larger than the energy range of accelerators which is about 10^2 GeV; perhaps such a large energy is meaningless.

We cannot, however, right away say that all of this is nonsense. In fact, Maxwell's equations correctly describe electromagnetic waves whose wavelengths vary from several hundred kilometers to 10^{-13} cm. The Newtonian mechanics certainly accurately describes both everyday phenomena and movements of celestial bodies; Einstein's theory of gravity is applicable to the entire Universe — about 10^{27} cm in length.

This "unification" energy of 10^{15} GeV is incredibly large as the energy of a single particle but it is not very large in comparison with the everyday scale; the energy corresponds to some 10^{-9} grams. Furthermore, a very much larger number is already included in the conventional theories. This is what is called Planck mass and indicates the energy range where the gravitational field becomes as important as the other fields, and where gravity must be treated quantum mechanically. This number is 10^{19} GeV, or 10^{-5} grams, so it is 10^4 times as large as the strong, electromagnetic, and weak unification energy.

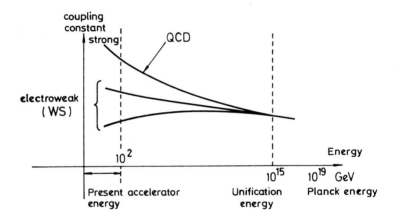

Fig. 20. 1 Unification energy for electromagnetic, weak and strong fields.

At the time when the renormalization was introduced in QED, Landau already noted the relationship of this theory to the gravitational field and pointed out that if there are a dozen elementary particles like the electron or the muon, the electromagnetic field and the gravitational field will be unified at the Planck energy. (The degree of change of the coupling constant differs depending on how many elementary particles there are, so the unification energy changes also. It is impossible, in any case, to obtain accurate numbers.)

Einstein's Unfulfilled Dream

The unification theories have a relatively long history; the gravitational field theory of Einstein may be said to be the beginning of this history. In Einstein's thinking, gravity is nothing other than that which reflects the geometrical characteristics of space. The reason material particles do not travel in straight lines around celestial objects is that the space itself is warped. Going straight on the surface of a sphere, like the earth, will bring us back to the original point; but we will never come back if the plane is truly flat. The difference between the two cases is due to the difference of the geometrical structure between the surface of a sphere and a truly flat surface, and it would be unnatural to say that there is a force on the objects on the surface of the sphere. According to Einstein, gravity is a similar phenomenon.

The only known long range force other than gravity is the electromagnetic force; this has not changed since Einstein's days. If gravity can be attributed to geometry, why can electromagnetism not be treated in the same way? This kind of thinking naturally occurs; the famous mathematician Hermann Weyl once tried to unify gravity and electromagnetism by expanding the Einstein's geometrical characteristics of space-time. His efforts were not successful.

Einstein himself also tried, in his last years, to formulate a unified field theory, but he never attained a satisfactory result. The idea of a unified field, however, did not die, and the seeds planted by Weyl have recently sprouted abundantly. This, though the place of application is different, is the idea of gauge fields; in fact, the name gauge was invented by Weyl.

The Unification of Three Forces and the Unification of Particles

The forces unified by the Weinberg-Salam theory are not gravity and electromagnetism but electromagnetism and the weak force. The next stage is an attempt to include the strong interaction as well; this is the idea presented at the beginning of this chapter. The unification theories of the weak, electromagnetic, and strong forces are called Grand Unified Theories, or GUTs.

Gravity, unfortunately, is not included in the Grand Unification. Incorporation of gravity must await the advent of some new theory in the future called, perhaps, Grand Grand Unified Theory. Of course, there are a lot of efforts being made toward such a theory; however, a grave obstacle in these efforts is that the mathematical characteristics of the gravitational field are very much different from those of the usual gauge theories, so that, for one, the renormalization is not possible when this field is treated quantum mechanically. Thus, for the gravitational field, we are still at the same stage as the theory of electrodynamics was before the advent of renormalization theory.

Even as a practical problem, since the quantum mechanical effects of gravity only manifest themselves in the region of the aforementioned Planck energy, there is no way to verify these unified theories experimentally. In any case, the immediate problem is to unify the forces other than the gravitational force — weak, strong, and electromagnetic — and at the same time to attempt the unification of matter particles such as quarks and leptons. I will discuss this in more detail in the next chapter.

21

PROGRAM FOR GRAND UNIFICATION

The Last Question

Einstein's equation of the gravitational field indicates that the gravitational field is determined by the distribution of matter, and is written thus:

$$R_{\mu\nu} - \tfrac{1}{2}\, g_{\mu\nu} R \;=\; 8\pi G T_{\mu\nu}.$$

The beauty of this equation probably cannot be appreciated without the knowledge of its mathematical meaning; but in a rough sense, the left-hand side represents the bending of space-time, in other words the gravitational field, and the right-hand side represents the energy in matter, that is to say particles, electromagnetic field, and all other kinds of energy that exist in space-time. G is Newton's gravitational constant.

Einstein's complaint.

Einstein complained once that the above equation is not very symmetrical. He said that, in analogy to architecture, the left-hand side is a beautiful building made of marble, but the right-hand side is a wooden shack. The reason for his complaint is that the gravitational field on the left-hand side is based on a beautifully simple geometrical principles, but the right-hand side seems to be very complex and random.

We have learned, however, that the electromagnetic field part of the right-hand side is an exception because, as a gauge field, it is an expression of a kind of geometrical principle. Furthermore, through the WS theory and QCD we can now include the weak and strong forces as gauge fields. What are left over are the matter particles, i.e., the leptons and quarks. Naturally, a Grand Unified Theory must also unify these matter particles. But is there a geometrical principle that applies to matter particles?

There is not yet a conclusive answer to this last question. There is a theory called supersymmetry which treats the fermions (the matter particles) and bosons (force fields) on the same frame-work; but the real problem is why matter particles possess mass. Each lepton or quark seems to have a mass which does not seem to be related to any other mass. This is precisely the reason why the right-hand side of Einstein's equation is complex. It is relatively easy to apply geometrical principles and supersymmetry principles to massless fields and particles.

Thus, Grand Unification is not something that derives a satisfactory answer to such fundamental questions as those posed above, but it is an attempt at a first step toward the unification of quarks, leptons, and others.

Let us now turn to a simple presentation of the Georgi-Glashow SU_5 theory as a representative example of Grand Unified Theories.

The Unification Theory of Georgi and Glashow

As we have seen many times, the difference between quarks and leptons lies in whether they interact strongly or not. To unify the strong interaction with the weak and electromagnetic interactions, the leptons and quarks, of course, must be unified. Then, the relationship between quarks and leptons can be thought of as something like that of charged and neutral particles.

We already know that the various flavors of leptons and quarks come in pairs like (u, d) and (ν, e). If the Cabibbo mixing (p. 185) is ignored, transformations between different pairs do not occur.

Let us then think of the lightest of the pairs of leptons and quarks, i.e., (u, d) and (ν, e), as a set. As was defined in p.190, these are the fundamental particles of the first generation, and the second generation is (c, s) and (ν_μ, μ). Since the difference between the generations concerns only mass, the unification that applies to the first generation should apply to all generations.

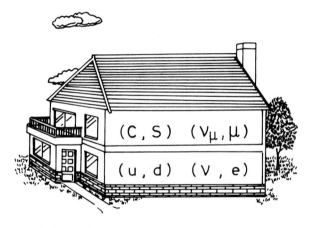

The first and second generations of quarks and leptons.

Considering the colors of quarks seems to bring the number of particles in the first generation up to 6 quarks and 2 leptons; but if we distinguish left- and right-handed particles, we see that the neutrino (the antineutrino) exists only in the left- (right-) handed version ν_L ($\bar{\nu}_L$). Even if the right-handed version exists, this will not interact through the weak interaction, as we presently know it. H. Georgi and S.L. Glashow, taking the above circumstances into consideration, introduced the SU_5 theory, which is described below.

Just as the SU_3 theory treated the u, d, s quarks as a group, the SU_5 theory starts with the 5 left-handed fundamental particles and the corresponding right-handed particles:

$(d_R, d_G, d_B, e, \nu)_L$;

$(d_R, d_G, d_B, e, \nu)_R$.

What about the rest of the particles? We need to wait a little longer for the answer to this question.

If these five particles are considered on completely the same footing, they have a SU_5 symmetry and free transformations among the particles are permitted. If we, however, divide the group into the first 3 and the last 2 and restrict transformations to be within each of these two subgroups, the symmetry becomes SU_3 and SU_2 respectively.

As can be seen from the above, the first kind of transformation changes the color of the d quark, and the latter the flavor of leptons; thus, they can be thought of as representing the strong and weak interactions. Note that the $\bar{d}_L$ and d_R do not participate in the weak interaction, as required by the V-A theory. So the SU_3 of color and SU_2 in the WS theory are incorporated in this theory.

The U_1 in the WS theory is thought to correspond to the quantum number that distinguishes the quarks (the first 3) and leptons (the last 2). The average electric charge of a group, as can be seen below, is zero.

$(\frac{1}{3}, \frac{1}{3}, \frac{1}{3}, -1, 0)_L$

$(-\frac{1}{3}, -\frac{1}{3}, -\frac{1}{3}, 1, 0)_R$

(This is the same characteristic as was used in the quark theory of Gell-Mann and Zweig.)

Are there Particles at Lower Levels than Quarks and Leptons?

The job is not finished yet. What about the u and other quarks? The answer to this question is rather complex; however, roughly speaking, the rest of the particles behave like composites made of the above five particles as far as the quantum numbers are concerned. This is like the old Sakata model (p. 96).

In the Sakata model, the proton, neutron and the lambda were the fundamental particles carrying the SU_3 of color, and the other baryons, such as the sigma particle and the xi particle, were seen as compounds of the fundamental three.

It seems to be the standard method of theorists to choose a minimum number of particles, from the stand-point of symmetry (or quantum number), as the fundamental set and treat the other particles as compounds of these. But the Sakata model was replaced by the quark model; it was not correct to take the three out of the eight baryons as fundamental particles.

Maybe, then, history will be repeated in the case of SU_5. Perhaps the particles that make up the quintet are particles at a level lower than quarks and leptons, the particles that make up the quarks and leptons.

Such questions are inevitable, and theorists, already, are turning their imaginations in such directions. But we cannot know if this is the correct direction. One of the reasons is that, unlike the case with the hadrons, there is absolutely no evidence that quarks and leptons are composites having some size. Another thing we can do is to scrap the SU_5 and treat all of the first generation fundamental particles on an equal footing.

The main purpose of the SU_5 theory was to unify quarks and leptons. Here quarks and leptons are not fundamentally different; for some reason, the symmetry is spontaneously broken, and the quintet breaks up into 3 quarks and 2 leptons, and for "low energy" phenomena, quarks and leptons exhibit completely different behaviors. But this energy is only "low" in comparison with the unification energy, so the symmetry between quarks and leptons probably cannot be observed in the laboratory in the foreseeable future.

An Astonishing Conclusion — All Matter is Unstable

The SU_5 theory, however, already includes a very important conclusion. This is the fact that there can be transformations among quarks and leptons; otherwise there would be no reason to consider a quintet of particles. But quarks are the structural constituents of all baryons, or nuclei, and thus are the carriers of baryon number. It has been thought that baryon number is absolutely conserved just as electric charge is. The result of a spontaneous breaking of the SU_5 symmetry would be that the baryon number is not conserved. Atomic nuclei would turn into leptons, and atoms would disappear! For example, proton decays like

$$p \rightarrow \bar{e} + \gamma, \qquad p \rightarrow \bar{e} + \pi^0$$

might be possible.

This is a serious problem and cannot be ignored. The immediate reaction would be that the SU_5 theory is wrong. But it can be argued that the conservation of baryon number is not as sacred as the conservation of electric charge. The reason is that there is no Coulomb-type force (long range gauge field) that corresponds to the baryon number. Thus, there is no theoretical mechanism that guarantees the conservation of baryon number.

A crane lives a thousand years, a tortoise ten thousand, but a proton lives 10^{30} years.

There is no evidence for the decay of the proton; speaking more quantitatively, the proton lifetime is thought to be not less than 10^{30} years. Since this number is incredibly larger than the age of the Universe, which is currently believed to be 10^{10} years, there is no cause to worry about disappearance of matter, but the point is important as a question of principle. Furthermore, the theoretical calculations indicate that the lifetime of the proton may not be much longer than 10^{30} years, which would make possible, with a little effort, the experimental measurement of the lifetime; in recent years many countries have been carrying out experiments that look for proton decay.

The attempts at Grand Unification have a very short history, no conclusive theories have emerged, and many unsolved problems remain; but a distinguishing characteristic of all these theories is that the scale of unification energy is very large. Some people claim that the next generation of accelerators must have energies in the range of 10^{15} GeV, up from 10^3 GeV, to achieve fundamentally new phenomena, and the range in between is a featureless region like the Sahara desert. If this is true, then particle physics may reach a wall at the end of this century.

The Coming Together of Particle Theory and Cosmology

It is impossible to attain the energies of 10^{15} GeV in the laboratory, but physicist have thought of a way to skillfully use other phenomena; this is the connection with cosmology. Since cosmology is not the subject of this book, the explanations cannot be complete. But briefly speaking, according to the presently accepted view, the Universe began with the Big Bang about 10 billion years ago and is continuing to expand.

The equation that governs this expansion is the Einstein equation (p. 212), and the size of the Universe has changed from 10^{-33} cm to its present size, 10^{27} cm. The former number corresponds to the inverse Planck mass (p. 224) 10^{-19} GeV, so the energy (or temperature) of matter inside the Universe at its birth was very high. It is easy to understand that the characteristics of elementary particles and gauge fields affected the development of the Universe from its beginning.

For example, there is the question of why the matter in the Universe is made up of baryons (protons, neutrons, etc.) and not from anti-matter. The baryon number of the Universe is the incredible number of 10^{80}, but shouldn't it be zero from the symmetry of particle and antiparticle (C, the charge conjugation symmetry)? (The baryon number is not large compared to the number of photons in the Universe, which is 10^{88}; but the latter is not strange since for photon there is no distinction between particles and antiparticle.)

There is a possibility of solving this problem if the unification theory is applied. As was pointed out many times before, the weak interaction does not respect many conservation laws. It has been recently pointed out that the occurrence of an imbalance between baryons and antibaryons during the development of the Universe can be understood by considering the characteristics of CP violation and baryon number non-conservation (proton decay). decay).

The other interesting problem is that of neutrino mass. The neutrino mass is very small experimentally and is generally believed to be zero; but there is a lack of theoretical necessity for this, unlike the case of the quantum of a gauge field. Recently, there have been experimental indications that the mass of the neutrino may not be zero. If this is true, the Universe should be full of "invisible" mass.

The Mystery of Man Who Divines Matter and Universe in an Instant

We have seen that cosmology and particle physics are intimately related, and they support each other. This is very fine; but is there a vast desert region as indicated by the naive Grand Unification Theories? The present author thinks not. In spite of the progress of unified gauge theories, there are many unanswered questions; and according to the past experience, nature has, from time to time, shown us that she is richer and more complex than our expectations. The future will probably be likewise full of surprises.

However, what we should be astonished with may be the fact that the secrets of nature do come to light one after the other. It is a mysterious experience to reflect that we, who constitute a portion of matter, in what must be called an instant of time at ten billion years after the birth of the Universe, are beginning to find the laws that govern the Universe, to know its history and also to divine that matter itself is a temporary existence with a finite lifetime.

GLOSSARY

Abelian and Non-Abelian groups

$$2 + 3 = 3 + 2, \qquad 2 \times 3 = 3 \times 2.$$

But there is a difference between going forward and then turning right and turning right and then going forward. So the operations of addition and multiplication are commutative, but the operations of changing direction and going forward are non-commutative.

Groups that are made up of commutative elements only are called Abelian groups, the other ones are called non-Abelian groups. For example, rotation around one axis is an Abelian group, whereas rotation around all axes (three-dimensional rotation) are non-Abelian. It was pointed out in Chapter 8 that the angular momentum, in quantum mechanics, is a conserved quantity associated with rotation; it turns out that the electric charge carried by charged particles possess similar mathematical characteristics as those of the angular momentum for one direction of rotation. On the other hand, the charge that corresponds to the Yang-Mills field, or the isospin, has as its name indicates, similar characteristics to the angular momentum of rotations in three dimensions. This is the reason why electromagnetic fields are called Abelian gauge fields (Abelian fields) and Yang-Mills fields non-Abelian gauge fields (non-Abelian fields).

Compton Wave Length

Length, time, and mass are the three basic quantities of physics in terms of which most other quantities can be expressed. Using Einstein's relationship ($E = mc^2$) and the Einstein-de Broglie relationships between energy and time, and momentum and length, the above three quantities can be converted, one from the other, in terms of the combinations of the speed of light c and Planck's constant h.

The length calculated using the method above, from the particle mass m, is called Compton wavelength, and is given by the form h/mc. This length expresses the quantum mechanical spread associated with the particle. The Compton wavelength of the electron is $\frac{1}{137}$ of the size of the hydrogen atom nucleus, that is to say about 10^{-11} cm; that of the pion is about the size of nuclei, about 10^{-13} cm.

Einstein-de Broglie Relationships

The particle energy E and momentum p are proportional to the wave number k (the inverse of the wavelength) and the frequency ν, respectively, of the quantum mechanical wave (wave function) that represents the particle. This is expressed by:

$$E = h\nu, \qquad p = hk.$$

The first equation was derived by Einstein and the latter by de Broglie. h is the Planck's constant. For reasons of convenience, $h/2\pi$ is often used and is denoted by $\hbar$. The value of $\hbar$ is 10^{-27} erg-sec.

Excited States

In general, composite particles (atoms, nuclei, hadrons, etc.) have many energy levels and can exist in any of these levels; the lowest energy level is called the ground state and all of the other levels are called excited states. Excited states are, in general, unstable and decay to the ground state by emission of particles, like photons, so the ground state is the normal state of the particle.

To make an excited state, a fixed energy must be given to the ground state from the outside. One way to achieve this is to carry out the opposite of the decay process. For example, if photons strike a composite particle, the desired reaction occurs when the photon energy matches the needed energy (excitation energy). When the excited state subsequently decays with the emission of a photon, the state is observed as a resonance of the photon scattering.

Fluctuation

The word fluctuation indicates the fact that physical quantities do not have a fixed value but change randomly. The cause of the fluctuation may be some outside influences; in the case of quantum mechanics, however, the fluctuation is fundamental, and, except in special cases (the so-called eigenstates of that physical quantity), appear inevitably and cannot be removed.

Heisenberg's Uncertainty Principle

In quantum mechanics, it is impossible to precisely measure the position x of a particle and its momentum p at the same time. Calling the uncertainty (fluctuation) of each quantity as Δx and Δp, the product of these uncertainties cannot be smaller than Planck's constant h. The same relationship holds between time and energy.

The origin of the uncertainty relationship is the wave nature of the particle. Waves possess an intrinsic width in space (or time) which increases inevitably when the wavelength is increased. Since the wavelength is inversely related to the momentum by the Einstein-de Broglie relationship, the above relationship holds.

If, for example, the momentum is precisely p, with Δp zero (thus Δx infinity), the state is an eigenstate of momentum and is said to possess eigenvalue p.

Lamb Shift

The phenomenon of the slight deviation of the energy levels of the electrons in the atom from those calculated from the Dirac equation is called Lamb shift. This phenomenon was discovered by the Columbia University Professor W. Lamb and became the first test to verify the correctness of the renormalization theory which was, at that time, in the process of being completed. The cause of the Lamb shift is the recoil of the electron and the consequent change of its energy levels, due to the fact that the electron is always emitting and reabsorbing virtual photons. Lamb received the Nobel Prize for his work.

Planck Mass (Planck Energy)

The gravitational force is very weak compared to electromagnetic and weak forces and can usually be ignored in the world of elementary particles. However, at very high energies, or very short distances, gravity must be treated quantum mechanically as well, and its effect cannot be ignored. The rough estimate of this energy is given by the Planck mass (energy). This is a combination of Newton's gravitational constant G, the speed of light c, and Planck's constant h, and is expressed by $(hc/G)^2$. The Planck mass is 2×10^{-5} grams, or 10^{19} GeV in energy. In terms of length (the Compton wavelength) or time, it is 10^{-33} cm or 10^{-44} seconds. These are called Planck length and Planck time respectively.

INDEX